Microstructural and Mechanical Characterization of Alloys

Microstructural and Mechanical Characterization of Alloys

Editors

Marek Sroka
Grzegorz Golański

MDPI • Basel • Beijing • Wuhan • Barcelona • Belgrade • Manchester • Tokyo • Cluj • Tianjin

Editors
Marek Sroka
Silesian University of Technology
Poland

Grzegorz Golański
Czestochowa University of Technology
Poland

Editorial Office
MDPI
St. Alban-Anlage 66
4052 Basel, Switzerland

This is a reprint of articles from the Special Issue published online in the open access journal *Crystals* (ISSN 2073-4352) (available at: https://www.mdpi.com/journal/crystals/special_issues/ Microstructural_Alloys).

For citation purposes, cite each article independently as indicated on the article page online and as indicated below:

LastName, A.A.; LastName, B.B.; LastName, C.C. Article Title. *Journal Name* **Year**, *Volume Number*, Page Range.

ISBN 978-3-03943-755-9 (Hbk)
ISBN 978-3-03943-756-6 (PDF)

Contents

About the Editors

Marek Sroka is Assistant Professor in the Department of Engineering Materials and Biomaterials at the Silesian University of Technology in Gliwice, Poland. Throughout his scientific career, he has participated in and organized numerous international scientific conferences. His scientific interests include materials science, materials for service at elevated temperatures, high-temperature creep resistance, creep tests, and computer aid in material engineering. He is author and coauthor of ca. 100 international scientific publications, including more than 40 publications in the Philadelphia list, and has won 20 awards and honors, both national and international, and is and/or has served as a contractor of more than 10 research and didactic projects in Poland and abroad, in addition to being a reviewer of numerous scientific publications.

Grzegorz Golański, Ph.D., is Professor at the Department of Materials Engineering, Czestochowa University of Technology. He specializes in the study of structure, heat treatment, and properties of engineering materials, mainly creep-resistant alloys. He is the author and coauthor of about 250 scientific publications, including seven monographs and books, and more than 40 publications in the Philadelphia list. He is also the author and coauthor of over 100 research works and gives expert opinions for industries. He has reviewed numerous scientific publications and won 17 awards. He is called upon by industry to provide his expert opinions.

Preface to "Microstructural and Mechanical Characterization of Alloys"

This book is a collection of manuscripts with original and innovative research studies which cover the recent developments in alloys (engineering materials), methods for improvement of the strength and cyclic properties of alloys, the stability of microstructure, the possible application of new (or improved) alloys, and the use of treatment for alloy improvement.

Metals and their alloys are currently the basic construction materials used in various fields of technology. The functional properties of these metallic materials depend on their chemical composition, (micro)structure, and production technology. Optimizing the functional properties of materials used in constructions to reduce their weight and increase the safety of use is currently the primary goal of engineers. This is done not only by introducing new types of materials with a better combination of properties but also by modifying the composition not only chemically but also through heat, thermomechanical, and thermochemical treatment.

The book aims to provide readers, students, and Ph.D. students as well as research personnel and professional engineers with information on the mechanical and physical properties decisive in the common use of metals and alloys as materials for construction, tools, or specific purposes. Additionally, this collection puts in one place not only theoretical studies of the recent developments in alloys but also the practical application of the discussed methods on particular examples and technological solutions. The manuscripts of this book are developed by renowned and respected researchers and specialists from around the world.

Marek Sroka, Grzegorz Golański
Editors

Editorial

Microstructural and Mechanical Characterization of Alloys

Marek Sroka [1],* and Grzegorz Golański [2]

[1] Department of Engineering Materials and Biomaterials, Silesian University of Technology, Konarskiego St. 18a, 44-100 Gliwice, Poland

[2] Department of Materials Engineering, Czestochowa University of Technology, Armii Krajowej 19, 42-200 Częstochowa, Poland; grzegorz.golanski@pcz.pl

* Correspondence: marek.sroka@polsl.pl; Tel.: +48-32-237-18-47

Received: 13 October 2020; Accepted: 13 October 2020; Published: 17 October 2020

Abstract: This Special Issue on "Microstructural and Mechanical Characterization of Alloys" features eight papers that cover the recent developments in alloys (engineering materials), methods of improvement of strength and cyclic properties of alloys, the stability of microstructure, the possible application of new (or improved) alloys, and the use of treatment for alloy improvement.

Keywords: metallic alloys; chemical composition; microstructure; treatment; mechanical properties

---

Metals and their alloys are currently the basic construction materials used in various fields of technology. The functional properties of these metallic materials depend on their chemical composition, (micro)structure and production technology. Optimizing the functional properties of materials used in construction to reduce their weight and increase the safety of use is currently the primary goal of engineers. This is achieved not only by introducing new types of materials with a better combination of properties, but also by modifying the chemical composition as well as heat, thermo-mechanical and thermo-chemical treatment.

The process of the engineering alloy's microstructure modification takes place not only through conventional plastic forming processes but also modern, unconventional methods, e.g., equal channel (micro)angular pressing (ECAP). In the case of the copper-based alloy Cu-0.43Mg [1], the ECAP process applied contributed to significant hardness increase, and the lower hardness region appeared at the area nearby the bottom surface. With the number of ECAP passes, the hardness gently increased and finally became saturated. The yield strength of the alloy increased from 124 MPa before the ECAP process to 555 MPa after eight ECAP passes.

Additionally, in the case of rolling of magnesium alloy Mg-2Y-0.6Nd-0.6Zr [2], the ECAP process contributed to obtaining high strength and low plasticity after rolling. As the number of ECAP passes increased, the grain size of the alloy gradually reduced, and the texture of the basal plane gradually weakened. The ultimate tensile strength of the alloy first increased and then decreased, the yield strength steadily lowered, and the plasticity continuously increased. After four passes of ECAP, the average grain size decreased from 11.2 μm to 1.87 μm, and the alloy obtained excellent comprehensive mechanical properties.

Another hardening mechanism for metallic alloys significantly influencing the increase in strength properties is the dispersion hardening and precipitation hardening. Strengthening the AA6063 aluminum alloy with fly ash (FA) particles and the production of AA6063-FA composite, as shown by research [3], leads to an increase in wear rate with increasing load, time and sliding velocity and the friction coefficient decreased with increasing these parameters. In the case of AZ91 magnesium alloy aged at different temperatures (T_a = 100 to 300 °C) for different durations (t_a = 4 to 192 h), the strengthening process was carried out through intermetallic β-Mg17Al12 phase-separated in the matrix α-Mg [4]. At lower ageing temperatures (100 and 150 °C) in the microstructure, only discontinuous precipitates were observed, while continuous precipitates invaded and formed at a high ageing temperature (300 °C). In regard to the ageing process, with the time at various ageing temperatures, magnesium alloy also contributed to the change in the crystalline lattice parameter ratio.

The effectiveness of the interaction of secondary (strengthening) phase particles depends not only on their size and distribution in the matrix but also on their thermodynamic stability. Increase in stability (low coarseness rate of the particles) strengthening fine molybdenum carbides precipitate in novel 5Cr5Mo2 steel during tempering treatment compared to H13 steel was observed [5]. Moreover, owing to their pinning effect on the dislocation slip, the dislocation density of the 5Cr5Mo2 steel decreases more slowly than that of the H13 steel. A slowdown of matrix softening of the steel processes can be obtained by modifying and optimizing the chemical composition and parameters of the heat treatment of the tested steel.

The softening process leads to a decrease in the strength properties of the alloys by reducing the strengthening mechanisms. Understanding the basic phenomena and processes related to the softening mechanism occurring in materials is not only the domain of steel but also such processes are observed in other groups of construction materials. In work [6], the study on the Al-Cu-Li-Mg-Ag alloy was performed. The alloy was deformed in a temperature range of 350–470 °C and a strain rate range of 0.01–10 s^{-1}. It has been shown that the main softening mechanism of this alloy is dynamic recovery. The conducted research allowed for the development of hot processing maps, which will enable the optimal selection of temperature and deformation parameters of the aluminum alloy.

Another way of increasing the functional properties of alloys is the fragmentation of the microstructure—reducing the grain size, reducing the width of the martensite/bainite strips, which allows one to obtain not only better strength, but also plastic properties. By modifying the austenitizing parameters of medium-carbon low-alloy martensitic steels, as shown in the studies presented in [7], it is possible to obtain steels with different lath martensite microstructures. It has been shown that with increasing the austenitizing temperature, the prior austenite grain size and block size increased, while the lath width decreased. Further, the yield strength and tensile strength increased due to the enhancement of the grain boundary strengthening.

Powder metallurgy is an alternative method of obtaining finished details and elements of machine or equipment parts. In the case of wear resistance of Co-based alloys with low carbon content [8], the application of the Selective Laser Sintering (SLS) and Powder Injection Molding manufacturing technique allowed us to obtain a product characterized by high properties, such as resistance to abrasive wear. The better resistance to abrasive wear for SLS was explained by the presence of a hard, intermetallic phase, present as precipitates limited in size and evenly distributed in the cobalt matrix and the structure of the cobalt matrix, with dominant content of the hexagonal phase.

This Special Issue covers very different aspects of microstructural–mechanical property relations identified by a wide variety of research techniques and their application in crystal engineering and material science.

The broad spectrum of topics included in the articles in this Special Issue shows that the microstructural and mechanical characteristics of alloy research are very modern. They are also of interest to scientists in other research centers [9–12], showing the long-term effects of temperature and time, as well as stresses on changes in the microstructure and the mechanical characterization of these materials, and that we can still expect new developments in this investigation field.

Conflicts of Interest: The authors declare no conflict of interest.

References

1. Ma, M.; Zhang, X.; Li, Z.; Xiao, Z.; Jiang, H.; Xia, Z.; Huang, H. Effect of Equal Channel Angular Pressing on Microstructure and Mechanical Properties of a Cu-Mg Alloy. *Crystals* **2020**, *10*, 426. [CrossRef]
2. Shi, X.; Li, W.; Hu, W.; Tan, Y.; Zhang, Z.; Tian, L. Effect of ECAP on the Microstructure and Mechanical Properties of a Rolled Mg-2Y-0.6Nd-0.6Zr Magnesium Alloy. *Crystals* **2019**, *9*, 586. [CrossRef]
3. Mohammed Razzaq, A.; Majid, D.L.; Ishak, M.R.; Muwafaq Basheer, U. Mathematical Modeling and Analysis of Tribological Properties of AA6063 Aluminum Alloy Reinforced with Fly Ash by Using Response Surface Methodology. *Crystals* **2020**, *10*, 403. [CrossRef]
4. Abd El-Rehim, A.F.; Zahran, H.Y.; Habashy, D.M.; Al-Masoud, H.M. Simulation and Prediction of the Vickers Hardness of AZ91 Magnesium Alloy Using Artificial Neural Network Model. *Crystals* **2020**, *10*, 290. [CrossRef]
5. Du, N.; Liu, H.; Fu, P.; Liu, H.; Sun, C.; Cao, Y.; Li, D. Microstructural Stability and Softening Resistance of a Novel Hot-Work Die Steel. *Crystals* **2020**, *10*, 238. [CrossRef]
6. Cao, L.; Liao, B.; Wu, X.; Li, C.; Huang, G.; Cheng, N. Hot Deformation Behavior and Microstructure Characterization of an Al-Cu-Li-Mg-Ag Alloy. *Crystals* **2020**, *10*, 416. [CrossRef]
7. Sun, C.; Fu, P.; Liu, H.; Liu, H.; Du, N.; Cao, Y. The Effect of Lath Martensite Microstructures on the Strength of Medium-Carbon Low-Alloy Steel. *Crystals* **2020**, *10*, 232. [CrossRef]
8. Ziębowicz, A.; Matus, K.; Pakieła, W.; Matula, G.; Pawlyta, M. Comparison of the Crystal Structure and Wear Resistance of Co-Based Alloys with Low Carbon Content Manufactured by Selective Laser Sintering and Powder Injection Molding. *Crystals* **2020**, *10*, 197. [CrossRef]
9. Golański, G.; Zielińska-Lipiec, A.; Zieliński, A.; Sroka, M.; Słania, J. Effect of long-term service on microstructure and mechanical properties of martensitic 9%Cr steel. *J. Mater. Eng. Perform.* **2017**, *26*, 1101. [CrossRef]
10. Zieliński, A.; Miczka, M.; Sroka, M. The effect of temperature on the changes of precipitates in low-alloy steel. *Mater. Sci. Technol.* **2016**, *32*, 1899. [CrossRef]
11. Golański, G.; Zieliński, A.; Sroka, M.; Słania, J. The Effect of Service on Microstructure and Mechanical Properties of HR3C Heat-Resistant Austenitic Stainless Steel. *Materials* **2020**, *13*, 1297. [CrossRef] [PubMed]
12. Zieliński, A.; Sroka, M.; Dudziak, T. Microstructure and Mechanical Properties of Inconel 740H after Long-Term Service. *Materials* **2018**, *11*, 2130. [CrossRef] [PubMed]

Publisher's Note: MDPI stays neutral with regard to jurisdictional claims in published maps and institutional affiliations.

Article

Effect of ECAP on the Microstructure and Mechanical Properties of a Rolled Mg-2Y-0.6Nd-0.6Zr Magnesium Alloy

Xiaofang Shi [1], Wei Li [1,2,*], Weiwei Hu [1], Yun Tan [1], Zhenglai Zhang [2] and Liang Tian [3]

[1] College of Material and Metallurgy, Guizhou University, Guiyang 550025, China;
    18285117265@163.com (X.S.); m18786670951@163.com (W.H.); tyl19941020@163.com (Y.T.)
[2] Zhejiang Huashuo Technology Co., Ltd., Ningbo 315000, China; 522321a02@sina.com
[3] Guizhou Province Technology Innovation Service Center, Guiyang 550004, China; 13017461042@163.com
* Correspondence: wli1@gzu.edu.cn; Tel.: +86-18085088978

Received: 14 October 2019; Accepted: 7 November 2019; Published: 8 November 2019

Abstract: A fine-grained Mg-2Y-0.6Nd-0.6Zr alloy was processed by bar-rolling and equal-channel angular pressing (ECAP). The effect of ECAP on the microstructure and mechanical properties of rolled Mg-2Y-0.6Nd-0.6Zr alloy was investigated by optical microscopy, scanning electron microscopy, electron backscattered diffraction and a room temperature tensile test. The results show that the Mg-2Y-0.6Nd-0.6Zr alloy obtained high strength and poor plasticity after rolling. As the number of ECAP passes increased, the grain size of the alloy gradually reduced and the texture of the basal plane gradually weakened. The ultimate tensile strength of the alloy first increased and then decreased, the yield strength gradually decreased, and the plasticity continuously increased. After four passes of ECAP, the average grain size decreased from 11.2 μm to 1.87 μm, and the alloy obtained excellent comprehensive mechanical properties. Its strength was slightly reduced compared to the as-rolled alloy, but the plasticity was greatly increased.

Keywords: magnesium alloy; ECAP; texture; mechanical properties

---

1. Introduction

As lightweight metallic materials used for engineering applications, magnesium alloys have the advantages of low density, high specific strength, high specific stiffness, good shielding and ease of recycling. They are widely used in numerous important areas, such as military, aerospace, transportation, and electronic communications [1–3]. Mg-Y-Nd-Zr (WE) alloys are commercial high-strength rare earth magnesium alloys developed in Britain in the 1980s. They have excellent creep resistance at high temperatures and are widely used as high-strength heat-resistant engineering materials [4]. However, their application potential is limited by their low number of slip systems and their poor plasticity at room temperature. In recent years, equal-channel angular pressing (ECAP) has been widely used as a method to effectively refine grains and improve the mechanical properties of magnesium alloys [5–8].

However, under conventional conditions, the ECAP of magnesium alloys can only be carried out at higher temperatures, which leads to grain growth during the pressing process and decreases the strengthening effect of ECAP. In this context, many scholars have begun to explore ways to reduce the pressing temperature, such as through the stepwise reduction of pressing temperature [9], the reduction of pressing speed [10], the application of back pressure [11,12], and a bread jacket outside the sample [13]. These methods all reduce the pressing temperature to a certain extent and the strengthening effect of ECAP is enhanced. However, with the development of society, the requirements for materials are ever increasing and the limited strengthening effect of ECAP limits its further expansion in industrial applications. Therefore, getting rid of the single ECAP strengthening mode, combining ECAP with other strengthening methods, breaking through the traditional ECAP strengthening limit and preparing fine-grained magnesium alloys with excellent performance have become hot issues in current ECAP research. At present, the most popular method is pre-deformation before ECAP. Pre-deformation can reduce the grain size, improve the as-casted microstructure, and enhance the plastic deformation ability of the alloy, thereby effectively reducing the ECAP temperature. In addition, pre-deformation can increase the strength of the alloy. This strengthening combined with ECAP strengthening can further improve the properties of the alloy. Miyahara et al. [14] first extruded an as-cast AZ61 magnesium alloy at 437 °C, and then conducted ECAP at 200 °C. After one pass of ECAP, a submicron microstructure was obtained and the average grain size after the fourth pass was ~0.62 μm. The elongation reached 1320% in the tensile test of strain rate of 3.3×10^{-4} s^{-1} at 200 °C. Krajňák et al. [15] first extruded an as-cast AX41 magnesium alloy at 350 °C and then ECAP was carried out at 220 °C and 250 °C. After eight passes of ECAP, both obtained good plasticity; however, after ECAP at 250 °C, the average grain size was larger than that after ECAP at 220 °C, the dislocation density was lower, and the texture was not conducive to the activation of the basal plane slip systems. These factors caused the yield strength after ECAP at 250 °C to be significantly lower than that after ECAP at 220 °C. Joungsik et al. [16] studied the ECAP of a AZ31 magnesium alloy sheet. They found that after the AZ31 magnesium alloy was plate-rolled, the base surface formed a typical rolling texture, i.e., the base surface was parallel to the rolling surface, resulting in the mechanical properties of sheet showing strong anisotropy. After ECAP along the route D at 225 °C, the severe shear deformation reduced the grain size of the alloy and developed basal texture with tilted basal planes towards the pressing direction. Ultimately, the anisotropy of the mechanical properties was reduced and the hardening behavior was enhanced.

Currently, the strengthening method of extrusion or plate-rolling combined with ECAP is relatively mature, but research on magnesium alloy bar-rolling combined with ECAP has rarely been reported. In this paper, an as-cast Mg-2Y-0.6Nd-0.6Zr alloy was studied. Bar-rolling was conducted first at 400 °C and then ECAP was carried out at 340 °C. The effect of ECAP on the microstructure and properties of the rolled Mg-2Y-0.6Nd-0.6Zr alloy was investigated by microstructure observation and a mechanical properties test. The aim of this study was to fill the gap of research on magnesium alloy bar-rolling combined with ECAP and provide a theoretical basis and technical support for improving the properties of magnesium alloys.

2. Materials and Methods

A Mg-2Y-0.6Nd-0.6Zr alloy was smelted in a well-type resistance furnace (Shiyan Electric Furnace Works, Shanghai, China) and 99.9% pure magnesium (Yinguang Huasheng Magnesium Company, Shanxi, China) along with Mg-25% Y, Mg-25% Nd, and Mg-30% Zr master alloys (Xinglin Nonferrous Metals Material Co., Ltd., Shanxi, China) were used to prepare it. A quartz crucible containing pure magnesium was placed in a well-type electric resistance furnace and RJ-5 solvent (Hengfeng Chemical Co., Ltd., Henan, China), which was composed of 56% anhydrous carnallite, 30% BaCl2 and 14% CaF2, was used as the covering agent and the refining agent. The furnace was heated to 720 °C with a heating rate of 10 °C/min. After the pure magnesium was completely melted, the Mg-Y, Mg-Nd and Mg-Zr master alloys were sequentially added to the crucible and the temperature of the furnace was raised to 780 °C. The solution was stirred when the master alloys were completely melted and then the power of the furnace was turned off so that the temperature of the solution dropped as the temperature of the furnace dropped. The crucible was taken out of the furnace while the solution was lowered to 720 °C and the solution was cast into a preheated cylindrical metal mold whose size was Φ30 mm × 200 mm and then water-cooled. The cast billets were homogenized at 450 °C for 6 hours and then air-cooled. The homogenized samples were rolled on a F50-150 bar-rolling machine (Hong Feng Ji Xie, Zhejiang, China) for seven passes at 400 °C with a total strain of 0.46. The samples with the dimensions of Φ12 mm × 80 mm were machined from the as-rolled bars, and then the samples were subjected to ECAP via a mold constructed in the laboratory. The mold structure is shown in Figure 1 and the angles of Φ and Ψ were 120° and 30°, respectively. The samples were pressed from one to six passes with a pressing velocity of 0.4 mm/s via route BC, i.e., the samples were rotated by 90° in the same direction between consecutive passes [17]. Prior to each pass, a layer of graphite and engine oil was applied to the inner wall of the mold and the surface of the sample as a lubricant and the samples were preheated together with the mold at 340 °C for 10 min. After each ECAP pass, the samples were quickly placed in water for cooling.

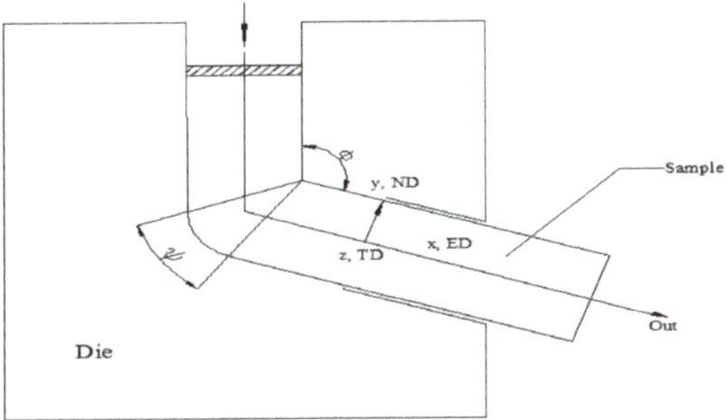

Figure 1. The schematic diagram of equal-channel angular pressing (ECAP) die (ED: extrusion direction, ND: normal direction, TD: transverse direction).

The microstructure of the samples was observed by a BH2 optical microscope (OM) (Olympus, Tokyo, Japan). Electron backscattered diffraction (EBSD) samples were prepared by a EM RES102 multi-function ion thinner (Leica, Wetzlar, Germany) and then observed the plane parallel to the extrusion direction (ED) or rolling direction (RD) on a S-3400N scanning electron microscope (Hitachi, Tokyo, Japan) and a NordlysMax3 electron backscatter diffractometer (Oxford, Abingdon, Britain) at an accelerating voltage of 20 kV and a step size of 0.2 μm. The EBSD data were analyzed by HKL Channel 5 software (Oxford, Abingdon, Britain) and the indexing rate reached 80%. The mechanical properties of the samples at room temperature were tested by an Instron 8501 universal tensile testing machine (Instron, Canton, USA). The dimensions of the tensile sample are shown in Figure 2, and were designed according to the standard of GB/T 228-2002, and the sampling direction was parallel to the ED. The tensile fracture morphology was analyzed on a SUPRA 40 scanning electron microscope (SEM) (Zeiss, Oberkochen, Germany).

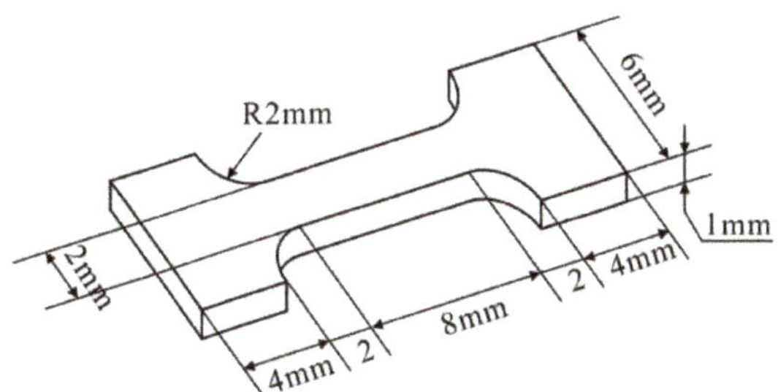

Figure 2. The dimensions of the tensile sample.

3. Results

3.1. Microstructure

Figures 3 and 4 display the microstructure and grain orientation distribution, respectively, of the Mg-2Y-0.6Nd-0.6Zr alloy subjected to rolling and after different numbers of ECAP passes. The grain size statistics and their distribution are shown in Figure 5. It can be seen in Figure 3; Figure 4 that after rolling, the grain size is relatively large, with individual large grains exceeding 30 μm and the average grain size being ~11.2 μm. After one pass of ECAP, the grain size of the alloy was remarkably reduced, some of the grains were elongated, the grain size presented a bimodal distribution, and the average grain size was ~2.43 μm. The fourth pass of ECAP led to fine equiaxed grains and the size distribution was concentrated in the range of 0–3 μm; however, at the same time, grains as large as 10 μm were also present and the average grain size was ~1.87 μm. After six passes of ECAP, the average grain size showed an increasing trend, the size of the coarse grains decreased, the grain size distribution was more homogeneous than that after four passes, and the average grain size was ~2.00 μm.

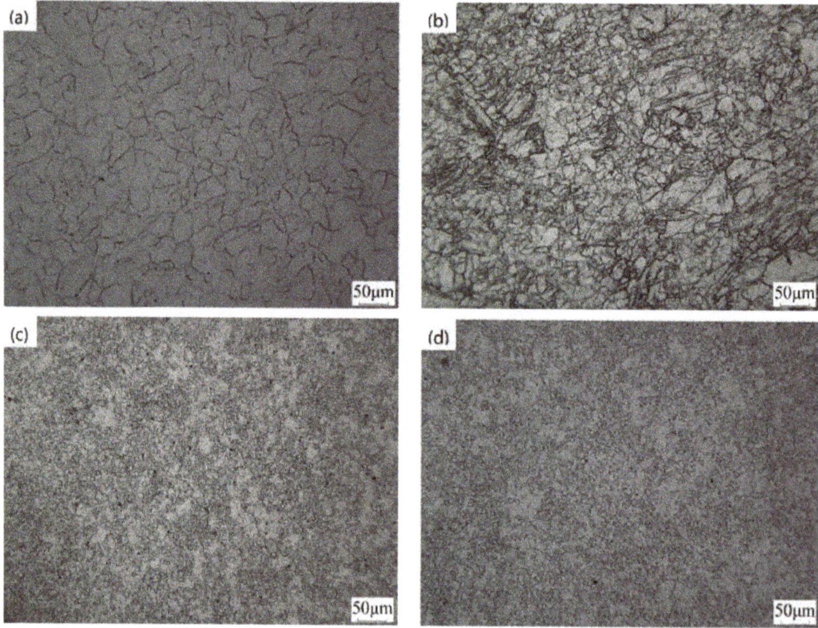

Figure 3. Microstructures of the Mg-2Y-0.6Nd-0.6Zr alloy. (**a**) as-rolled; (**b**) one pass; (**c**) four passes; (**d**) six passes.

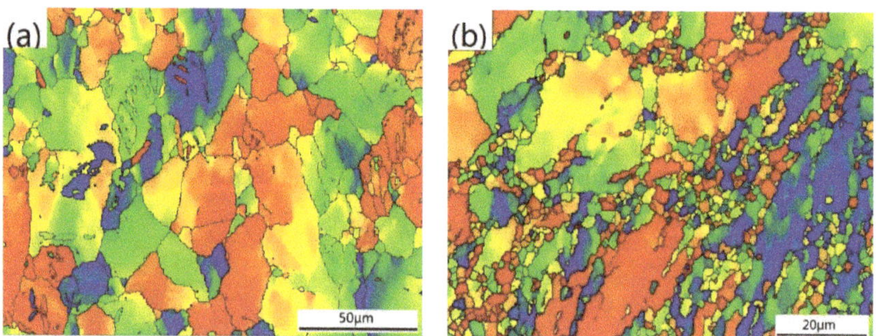

Figure 4. *Cont.*

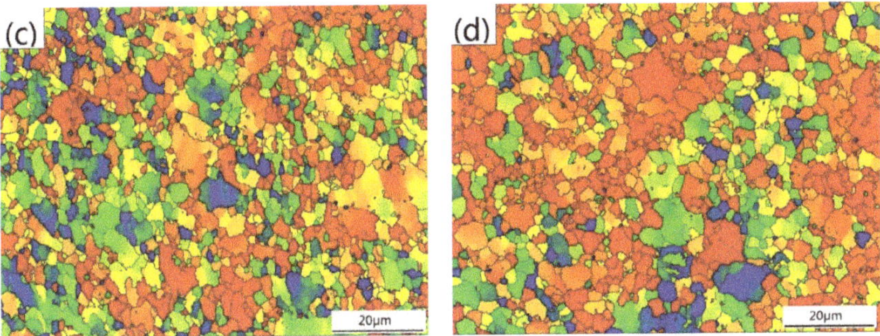

Figure 4. Grain orientation distribution of the Mg-2Y-0.6Nd-0.6Zr alloy. (**a**) as-rolled; (**b**) one pass; (**c**) four passes; (**d**) six passes.

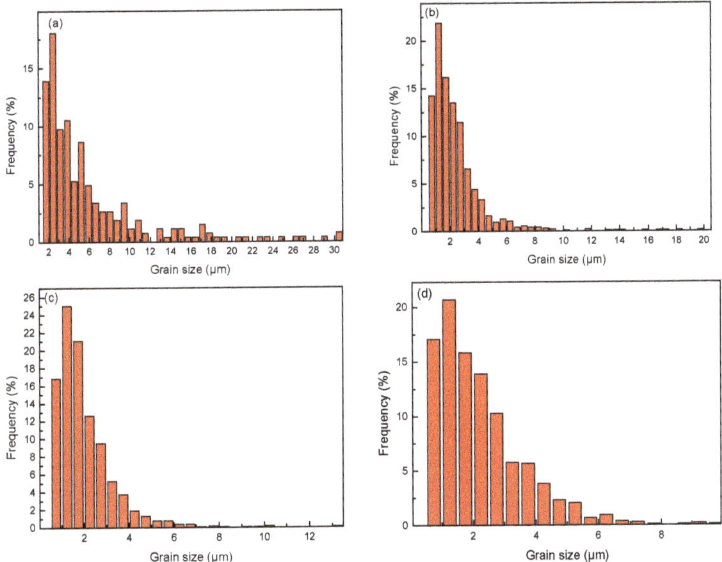

Figure 5. Grain size distribution of the Mg-2Y-0.6Nd-0.6Zr alloy. (**a**) as-rolled; (**b**) one pass; (**c**) four passes; (**d**) six passes.

3.2. Texture

Figure 6 presents the pole figure of the Mg-2Y-0.6Nd-0.6Zr alloy subjected to rolling and a different number of ECAP passes. The as-rolled alloy has a strong (0001) texture with the strongest pole density of 12.30, and the basal plane of most grains is nearly parallel to the rolling direction, as shown in Figure 6a. The pole figure of the alloy after different numbers of ECAP passes (Figure 6b–d), which indicates that the texture in the basal plane was rotated and the strength was continuously weakened with an increasing number of ECAP passes. After one pass of ECAP, the basal plane texture became dispersed, the basal plane of some grains was parallel to the extrusion direction, and the strongest pole density was decreased to 11.34. After four ECAP passes, the basal plane texture was rotated because the specimen rotated 90° along the same direction after each extrusion, the basal plane was ~30° from the extrusion direction, and the strongest pole density further decreased to 7.61. After six ECAP passes, the basal plane formed a typical inclined texture whose basal plane was parallel to the shear plane and was ~45° to the extrusion direction. The strongest pole density was 6.93, which is slightly lower than after four passes.

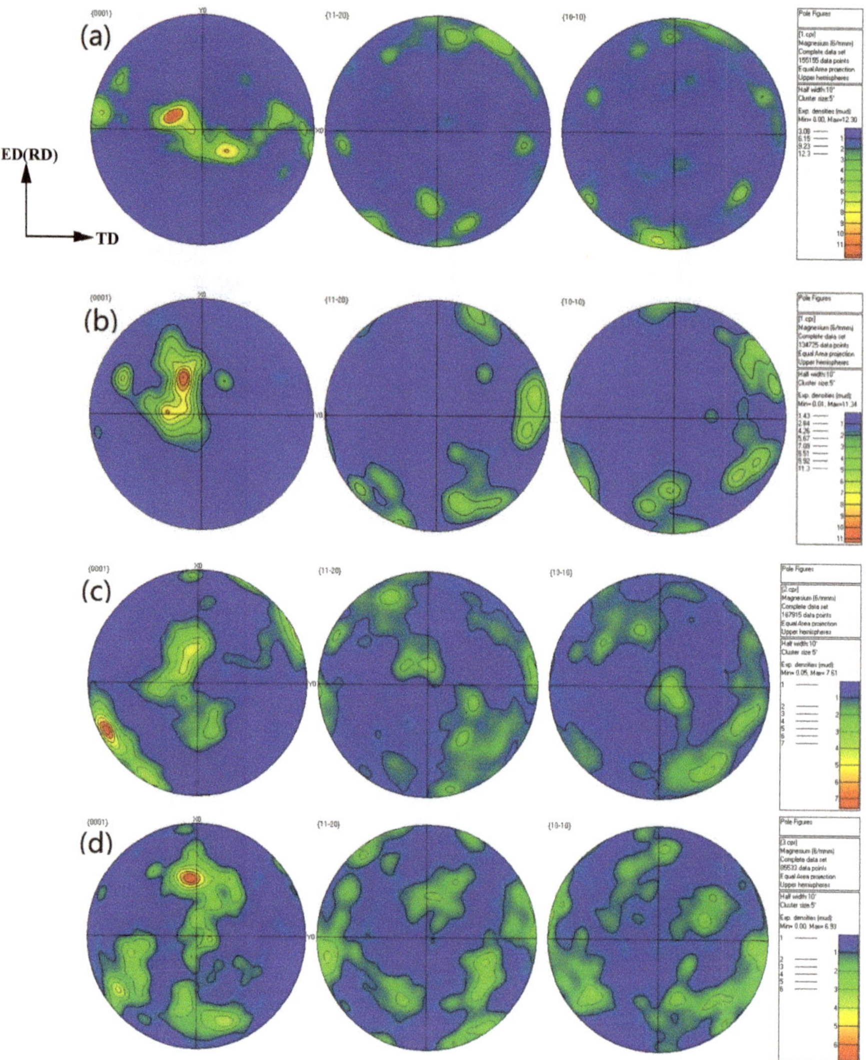

Figure 6. The pole diagram of the Mg-2Y-0.6Nd-0.6Zr alloy. (**a**) as-rolled; (**b**) one pass; (**c**) four passes; (**d**) six passes.

3.3. Mechanical Properties at Room Temperature

The samples after rolling and a different number of ECAP passes were subjected to a tensile test at room temperature. It can be seen in Figure 7 and Table 1 that the as-rolled alloy had the highest ultimate tensile strength and yield strength with values of 246 MPa and 216 MPa, respectively. However, the plasticity of the as-rolled alloy was extremely poor, and elongation was only 3.8%. With an increasing number of ECAP passes, the ultimate tensile strength first increased and then decreased, while the yield strength continuously decreased and the elongation continuously increased. After the first pass of ECAP, the ultimate tensile strength and yield strength were significantly reduced to 213 MPa and 182 MPa, respectively, but the plasticity was improved and the elongation increased to 12.3%. After four passes of ECAP, the ultimate tensile strength was greatly improved compared with the first pass of ECAP, and the yield strength did not change appreciably. The ultimate tensile strength and yield strength were 238 MPa and 180 MPa, respectively, and the elongation increased further to 19.7%. After six passes of ECAP, the alloy had the lowest ultimate tensile strength and yield strength with values of 209 MPa and 148 MPa, respectively, but the elongation reached 27.5%.

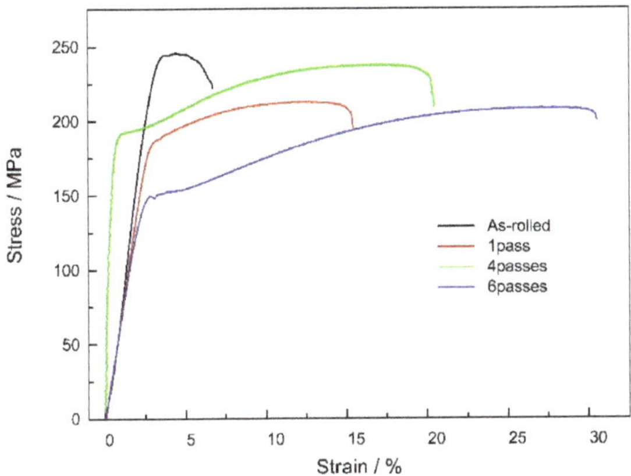

Figure 7. The tensile stress–strain curves of the Mg-2Y-0.6Nd-0.6Zr alloy.

Table 1. The mechanical properties of the Mg-2Y-0.6Nd-0.6Zr alloy.

State	Ultimate Tensile Strength/MPa	Yield Strength/MPa	Elongation/%
As-rolled	246 ± 8.3	216 ± 7.4	3.8 ± 0.12
One pass	213 ± 6.5	182 ± 5.3	12.3 ± 0.36
Four passes	238 ± 7.7	180 ± 6.2	19.7 ± 0.45
Six passes	209 ± 4.9	148 ± 5.5	27.5 ± 0.41

Figure 8 presents the tensile fracture morphology of the Mg-2Y-0.6Nd-0.6Zr alloy subjected to rolling and a different number of ECAP passes. The fracture of the as-rolled alloy (Figure 8a), is relatively flat and bright, which is a typical brittle fracture, indicating that the plasticity of the as-rolled alloy is poor. It can be seen in Figure 8b that after four passes of ECAP, a large number of dimples appeared on the fracture surface of the alloy, which indicates a typical ductile fracture, meaning that the plasticity of the alloy after ECAP was greatly improved compared with that of the as-rolled alloy. After six passes of ECAP, the dimples on the fracture surface were more uniform and deeper than those in the fourth pass, as shown in Figure 8c. This indicates that the plasticity of the alloy after six passes of ECAP is further increased compared with that after four passes of ECAP, which is consistent with the room temperature tensile test results.

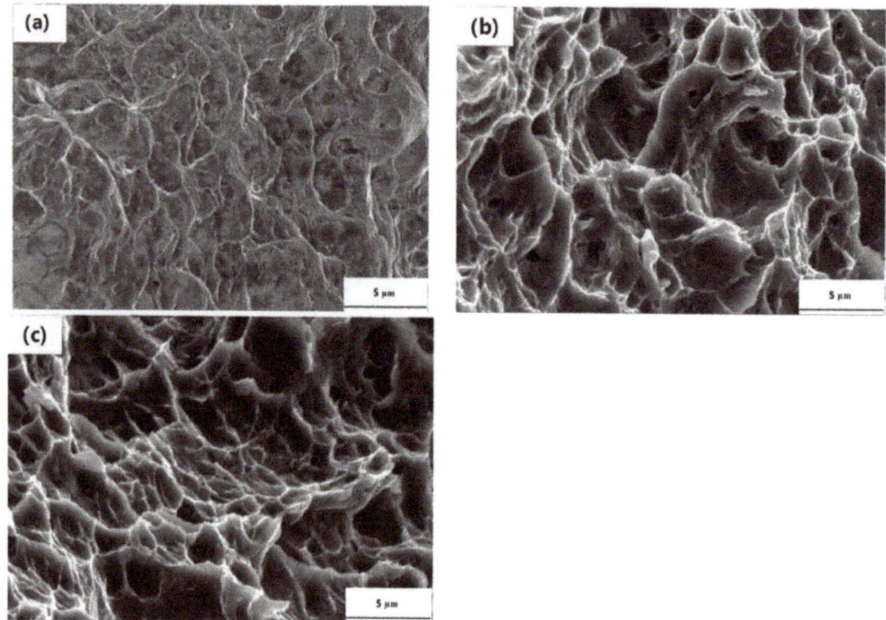

Figure 8. The tensile fracture morphology of the Mg-2Y-0.6Nd-0.6Zr alloy. (**a**) as-rolled; (**b**) four passes; (**c**) six passes.

4. Discussion

ECAP can effectively reduce the grain size of an alloy and its refining effect mainly depends on the processing temperature and total strain [18,19]. Magnesium alloys are refined by mechanical shear and dynamic recrystallization during ECAP [20]. The grains are twisted, sheared and broken when the magnesium alloy experiences shearing deformation in the mold so that the original coarse grains are divided into a plurality of fine grains; on the other hand, shear deformation can produce many dislocation tangles and shear bands which could provide nucleation sites and a driving force for dynamic recrystallization, which further reduces the grain size. It is well known that five independent slip systems are necessary for plastic deformation of polycrystalline materials. There are only two independent slip systems in the basal slip systems of magnesium alloys [21] and the non-basal slip systems of magnesium alloys are difficult to activate at low temperatures because the critical resolved shear stress (CRSS) of non-basal slip systems is much larger than that of basal slip systems [22]. Twinning is an important mechanism for low temperature plastic deformation of magnesium alloys because it can change crystal orientation and release the stress concentration caused by dislocation plugging. At high temperatures, the CRSS of the non-basal slip systems of magnesium alloys is greatly reduced and becomes easy to activate [23]. At that point, The plastic deformation of magnesium alloys does not depend on twinning but mainly depends on dislocation slipping [24]. In this study, both the rolling process and ECAP were carried out at relatively high temperatures, so twins would not be generated in the deformed microstructure, which is consistent with the results in Figures 3 and 4. In addition to the deformation temperature, the deformation mechanism of magnesium alloys is closely related to the grain size. The deformation mechanism of coarse grains is a typical slipping and twinning mechanism, while in the case of fine grains, in addition to slipping and twinning, grains can also coordinate deformation through grain boundary sliding and rotation [25]. These mechanisms work together to improve the deformation capacity of the alloys. When the Mg-2Y-0.6Nd-0.6Zr alloy was rolled at 400 °C, the grain size of the alloy was relatively large, the main deformation mechanism of the alloy was dislocation slipping, and the non-basal slip played an important role in order to satisfy the conditions of the five independent slip systems. The seven-pass rolling treatment caused the dislocation to proliferate in the grains, forming a large number of dislocation tangles and storing a lot of energy. In the subsequent ECAP process, these dislocation tangles and the stored energy can provide favorable conditions for dynamic recrystallization. With an increasing number of ECAP passes, the grain size is continuously reduced under the combined action of mechanical shearing and dynamic recrystallization, the deformation mechanism of the alloy is transformed from solely dislocation slipping to a combination of dislocation slipping and grain boundary sliding and rotation, and the plastic deformation capacity of the alloy is continuously enhanced. After four passes of ECAP, the dynamic recrystallization of the alloy was almost complete, the effect of mechanical shearing and dynamic recrystallization was greatly weakened, and the grain size would have increased if the extrusion passes has been further increased. Feng et al. [26,27] also found in the ECAP studies of an AZ31 magnesium alloy and Al-Mg-Si alloy that the grain size has a tendency to grow after the ECAP reaches a certain number of passes.

It can be seen in Table 1 that the as-rolled alloy kept the highest strength and the worst plasticity. This is because the as-rolled alloy accumulated a large amount of strain after seven passes of rolling, which led to the increase in dislocation density and the increase in the plastic deformation resistance of the alloy. In addition, water-cooling produces greater internal stress in the alloy after rolling, which accelerates the crack extension rate. Finally, the as-rolled alloy obtained a high strength and a poor plasticity due to work-hardening. After one pass of ECAP, the average grain size of the alloy decreased from ~11.2 μm to ~2.43 μm. It is well known that the finer the grain size, the better the strength and plasticity of the alloy, but the strength of the alloy decreased after one pass of ECAP. This is due to the fact that the dislocations were annihilated during the process of thermal insulation and subsequent ECAP so that the work-hardening effect was substantially weakened compared with the as-rolled alloy. Finally, the strength of the alloy decreased and the plasticity increased after one pass of ECAP.

With an increasing number of ECAP passes, the grain size of the alloy was gradually reduced, and the elongation increased, but the yield strength gradually decreased. In particular, after the sixth pass of ECAP, the yield strength decreased from a value of 180 MPa for the fourth pass to 148 MPa, which is contrary to the traditional Hall–Petch relationship. Kim et al. [28] concluded that the mechanical properties of magnesium alloys are closely related to texture and grain size after ECAP. When the base plane is 45° from the extrusion direction, the Schmid factor (SF) tends to be 0.5, the alloy is in a soft orientation, and the yield strength decreases. When the base plane is parallel or perpendicular to the extrusion direction, the SF tends to be 0, the alloy is in a hard orientation, and the yield strength increases. Figure 9 presents the SF distribution of the Mg-2Y-0.6Nd-0.6Zr alloy after different number of ECAP passes. Figure 9a shows that the SF of the as-rolled alloy is low with an average value of 0.282. After one pass of ECAP, because the basal plane of some grains was parallel to the extrusion direction (Figure 6a), the ratio of SF approaching 0 was higher, but the average value, 0.285, was close to that of the as-rolled. At that time, the alloy was in a hard orientation so the yield strength was improved. After four passes of ECAP, the average grain size of the alloy decreased from ~2.43 μm to ~1.87 μm, while the strength and plasticity of the alloy increased. However, the average value of the SF increased from 0.285 to 0.317; the softening effect of texture is equivalent to the strengthening effect of grain refinement, so the yield strength did not change significantly. After six passes of ECAP, the alloy formed a typical inclined texture whose basal plane was parallel to the shear plane and was ~45° to the extrusion direction. Here, the ratio of SF factor approaching 0.5 was higher, with an average value of 0.354, and the alloy was in a soft orientation. Moreover, the grain size of the alloy after six passes of ECAP was slightly larger than that after four passes of ECAP, so the yield strength was greatly reduced compared with the fourth pass of ECAP. However, due to the uniform distribution of the grain size, the deformation compatibility of the alloy increased. Therefore, the plasticity of the alloy after six ECAP passes was higher than that after the fourth ECAP pass and the elongation reached

27.5%. Muralidhar et al. [29,30] also obtained similar conclusions in their ECAP studies of AZ31 and AZ80 magnesium alloys.

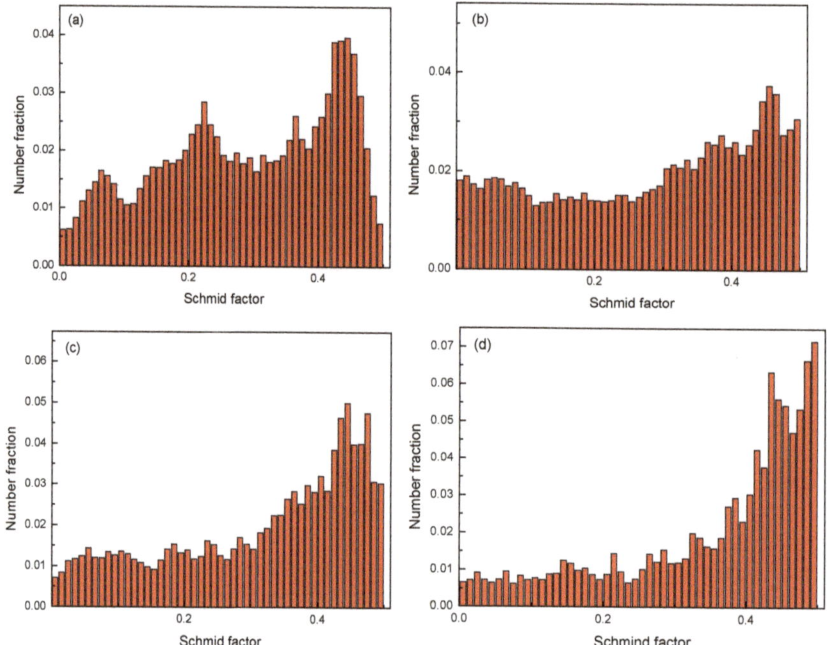

Figure 9. The Schmid factor distribution of the Mg-2Y-0.6Nd-0.6Zr alloy. (**a**) as-rolled; (**b**) one pass; (**c**) four passes; (**d**) six passes.

5. Conclusions

In this paper, the ECAP of a rolled Mg-2Y-0.6Nd-0.6Zr alloy was successfully processed at 340 °C. The microstructure and mechanical properties of the Mg-2Y-0.6Nd-0.6Zr alloy before and after ECAP were investigated. The following are the main conclusions:

(1) After the Mg-2Y-0.6Nd-0.6Zr alloy was rolled for seven passes at 400 °C, a high strength was obtained due to work-hardening. The ultimate tensile strength and yield strength were 216 MPa and 246 MPa, respectively, but the plasticity was extremely poor and the elongation was only 3.8%.

(2) ECAP of the Mg-2Y-0.6Nd-0.6Zr alloy was carried out at 340 °C. With an increasing number of ECAP passes, the grain size of the alloy gradually decreased under the combined action of mechanical shearing and dynamic recrystallization. After four passes of ECAP, the average grain size of the alloy decreased from 11.2 μm to 1.87 μm and the grain size no longer decreased as the number of ECAP passes increased.

(3) With an increasing number of ECAP passes, the plasticity of the Mg-2Y-0.6Nd-0.6Zr alloy increased continuously and the yield strength and tensile strength first increased and then decreased due to a combination of the fine-grain strengthening and texture softening. After four passes of ECAP, good comprehensive mechanical properties were obtained and the strength was slightly decreased compared with the as-rolled alloy, but the plasticity was greatly increased.

Author Contributions: W.L. and L.T. designed the experiment; Y.T. and Z.Z. conducted the experiment; testing, W.H.; X.S., collected data, analyzed and wrote the paper; checking and modifying, W.L.

Acknowledgments: This research was funded by the National Natural Science Foundation of China (51661007), China Postdoctoral Science Foundation (2018M633416), Guizhou Science and Technology Fund (qian ke heJ [2017] 1022), Guizhou Science and Technology Platform Project (qian ke he talent platform [2019]5053).

Conflicts of Interest: The authors declare no conflict of interest.

References

1. Aghion, E.; Bronfin, B.; Eliezer, D. The role of the magnesium industry in protecting the environment. *J. Mater. Proc. Technol.* **2001**, *117*, 381–385. [CrossRef]
2. Mordike, B.L.; Ebert, T. Magnesium: Properties—applications—potential. *Mater. Sci. Eng. A* **2001**, *302*, 37–45. [CrossRef]
3. Luo, A.A.; Mishra, R.K.; Powell, B.R.; Sachdev, A.K. Magnesium Alloy Development for Automotive Applications. *Mater. Sci. Forum* **2012**, *706*, 69–82. [CrossRef]
4. Wang, X.; Liu, C.; Xu, L.; Xiao, H.; Zheng, L. Microstructure and mechanical properties of the hot-rolled Mg–Y–Nd–Zr alloy. *J. Mater. Res.* **2013**, *28*, 1386–1393. [CrossRef]
5. Yamashita, A.; Horita, Z.; Langdon, T.G. Improving the mechanical properties of magnesium and a magnesium alloy through severe plastic deformation. *Mater. Sci. Eng. A* **2001**, *300*, 142–147. [CrossRef]
6. Chen, B.; Lin, D.L.; Jin, L.; Zeng, X.Q.; Lu, C. Equal-channel angular pressing of magnesium alloy AZ91 and its effects on microstructure and mechanical properties. *Mater. Sci. Eng. A* **2008**, *483*, 113–116. [CrossRef]
7. Ma, A.; Jiang, J.; Saito, N.; Shigematsu, I.; Yuan, Y.; Yang, D.; Nishida, Y. Improving both strength and ductility of a Mg alloy through a large number of ECAP passes. *Mater. Sci. Eng. A* **2009**, *513*, 122–127. [CrossRef]
8. Figueiredo, R.B.; Langdon, T.G. Grain refinement and mechanical behavior of a magnesium alloy processed by ECAP. *J. Mater. Sci.* **2010**, *45*, 4827–4836. [CrossRef]
9. Biswas, S.; Dhinwal, S.S.; Suwas, S. Room-temperature equal channel angular extrusion of pure magnesium. *Acta Mater.* **2010**, *58*, 3247–3261. [CrossRef]
10. Ding, S.X.; Lee, W.T.; Chang, C.P.; Chang, L.W.; Kao, P.W. Improvement of strength of magnesium alloy processed by equal channel angular extrusion. *Scr. Mater.* **2008**, *59*, 1006–1009. [CrossRef]
11. Xu, C.; Xia, K.; Langdon, T.G. Processing of a magnesium alloy by equal-channel angular pressing using a back-pressure. *Mater. Sci. Eng. A* **2009**, *527*, 205–211. [CrossRef]
12. Zhang, N.X.; Ding, H.; Li, J.Z.; Wu, X.L.; Li, Y.L.; Xia, K. Microstructure and Mechanical Properties of Ultra-Fine Grain AZ80 Alloy Processed by Back Pressure Equal Channel Angular Pressing. *Mater. Sci. Forum* **2011**, *667*, 547–552. [CrossRef]
13. Lei, W.; Wei, L.; Wang, H.; Sun, Y. Effect of annealing on the texture and mechanical properties of pure Mg by ECAP at room temperature. *Vacuum* **2017**, *144*, 281–285. [CrossRef]
14. Miyahara, Y.; Horita, Z.; Langdon, T.G. Exceptional superplasticity in an AZ61 magnesium alloy processed by extrusion and ECAP. *Mater. Sci. Eng. A* **2006**, *420*, 240–244. [CrossRef]
15. Krajňák, T.; Minárik, P.; Stráská, J.; Gubicza, J.; Máthis, K.; Janeček, M. Influence of equal channel angular pressing temperature on texture, microstructure and mechanical properties of extruded AX41 magnesium. *J. Alloys Compd.* **2017**, *705*, 273–282. [CrossRef]
16. Suh, J.; Victoria-Hernández, J.; Letzig, D.; Golle, R.; Volk, W. Enhanced mechanical behavior and reduced mechanical anisotropy of AZ31 Mg alloy sheet processed by ECAP. *Mater. Sci. Eng. A* **2016**, *650*, 523–529. [CrossRef]
17. Furukawa, M.; Iwahashi, Y.; Horita, Z.; Nemoto, M.; Langdon, T.G. The shearing characteristics associated with equal-channel angular pressing. *Mater. Sci. Eng. A* **1998**, *257*, 328–332. [CrossRef]
18. Akihiro, Y.; Daisuke, Y.; Zenji, H.; Terence, G. Langdon Influence of pressing temperature on microstructural development in equal-channel angular pressing. *Mater. Sci. Eng. A* **2000**, *287*, 100–106.
19. Ramin, J.; Mohammad, S.; Hamid, J. ECAP effect on the micro-structure and mechanical properties of AM30 magnesium alloy. *Mater. Sci. Eng. A* **2014**, *593*, 178–184.
20. Su, C.W.; Lu, L.; Lai, M.O. A model for the grain refinement mechanism in equal channel angular pressing of Mg alloy from microstructural studies. *Mater. Sci. Eng. A* **2006**, *434*, 227–236. [CrossRef]
21. Yoo, M.H. Slip, Twinning, and Fracture in Hexagonal Close-Packed Metals. *Metall. Trans. A* **1981**, *12*, 409–418. [CrossRef]
22. Koike, J. Enhanced deformation mechanisms by anisotropic plasticity in polycrystalline Mg alloys at room temperature. *Metall. Mater. Trans. A* **2005**, *36*, 1689–1696. [CrossRef]

23. Xin, R.L.; Wang, B.S.; Zhou, Z.; Huang, G.J.; Liu, Q. Effects of strain rate and temperature on microstructure and texture for AZ31 during uniaxial compression. *Trans. Nonferrous Met. Soc. China* **2010**, *20*, s594–s598. [CrossRef]

24. Chino, Y.; Kimura, K.; Mabuchi, M. Twinning behavior and deformation mechanisms of extruded AZ31 Mg alloy. *Mater. Sci. Eng. A* **2008**, *486*, 481–488. [CrossRef]

25. Partridge, P.G. The crystallography and deformation modes of hexagonal close-packed metals. *Metall. Rev.* **1967**, *12*, 169–194.

26. Feng, X.M.; Tao-Tao, A.I. Microstructure evolution and mechanical behavior of AZ31 Mg alloy processed by equal-channel angular pressing. *Trans. Nonferrous Met. Soc. China* **2009**, *19*, 293–298. [CrossRef]

27. Khelfa, T.; Rekik, M.A.; Khitouni, M.; Cabrera-Marrero, J.M. Structure and microstructure evolution of Al–Mg–Si alloy processed by equal-channel angular pressing. *Int. J. Adv. Manuf. Technol.* **2017**, *92*, 1731–1740. [CrossRef]

28. Kim, W.J.; Hong, S.I.; Kim, Y.S.; Min, S.H.; Jeong, H.T.; Lee, J.D. Texture development and its effect on mechanical properties of an AZ61 Mg alloy fabricated by equal channel angular pressing. *Acta Mater.* **2003**, *51*, 3293–3307. [CrossRef]

29. Muralidhar, A.; Narendranath, S.; Nayaka, H.S. Effect of equal channel angular pressing on AZ31 wrought magnesium alloys. *J. Magnes. Alloys* **2013**, *1*, 336–340. [CrossRef]

30. Wang, L.; Mostaed, E.; Cao, X.; Huang, G.; Fabrizi, A.; Bonollo, F.; Chi, C.; Vedani, M. Effects of texture and grain size on mechanical properties of AZ80 magnesium alloys at lower temperatures. *Mater. Des.* **2016**, *89*, 1–8. [CrossRef]

Article

Comparison of the Crystal Structure and Wear Resistance of Co-Based Alloys with Low Carbon Content Manufactured by Selective Laser Sintering and Powder Injection Molding

Anna Ziębowicz [1], Krzysztof Matus [2], Wojciech Pakieła [2], Grzegorz Matula [2] and Miroslawa Pawlyta [2,*]

[1] Department of Biomaterials and Medical Devices Engineering, Silesian University of Technology, Roosevelt 40 St., 41-800 Zabrze, Poland; anna.ziebowicz@polsl.pl
[2] Department of Engineering Materials and Biomaterials, Silesian University of Technology, Konarskiego 18A, 44-100 Gliwice, Poland; krzysztof.matus@polsl.pl (K.M.); wojciech.pakiela@polsl.pl (W.P.); grzegorz.matula@polsl.pl (G.M.)
* Correspondence: miroslawa.pawlyta@polsl.pl

Received: 31 December 2019; Accepted: 2 March 2020; Published: 13 March 2020

Abstract: Cobalt alloys are widely used in biomedicine, implantology, and dentistry due to their high corrosion resistance and good mechanical properties. The high carbon improves the wear properties, but causes fragility and dangerous cracking of elements during use. The aim of the present work was to analyze and compare the structure and wear resistance of Co-based alloy samples with low carbon content, produced by Selective Laser Sintering (SLS) and Powder Injection Molding (PIM). Structure characterization, mainly with the use of transmission electron microscopy, was applied to investigate the differences in tribological properties. The better resistance to abrasive wear for SLS was explained by the presence of a hard, intermetallic phase, present as precipitates limited in size and evenly distributed in the cobalt matrix. The second factor was the structure of the cobalt matrix, with dominant content of the hexagonal phase. By combining the characteristic features of the matrix and the reinforcing phase, the analyzed material gains an additional advantage, namely a higher resistance to abrasive wear.

Keywords: Co-based alloy; tribological properties; wear; microstructure; selective laser sintering (SLS); powder injection molding (PIM)

---

1. Introduction

The structure of a Co-based alloy consists of a low-temperature phase ε (hexagonal close-packed (hcp)) and a phase γ (face-centered cubic (fcc)), which is formed at a higher temperature. In equilibrium conditions, only the ε phase exists. However, the $\gamma \rightarrow \varepsilon$ transformation in Co and its alloys is very slow, due to the limited chemical driving forces available at the transformation temperature. Thus, especially at high cooling rates, the transformation is retained below the phase boundary in a metastable state [1,2]. It is also worth noting that the reverse transformation $\varepsilon \rightarrow \gamma$ is rare in Co-based alloys, and the ε phase is much more stable than the γ phase at room temperature [3]. The $\gamma \rightarrow \varepsilon$ transformation is associated with the decreasing value of the stacking fault energy [4], and the martensite ε phase is greatly known to reduce ductility due to the relatively lower number of effective slip systems [5]. Co-based alloys, especially for medical applications, contain added chromium and molybdenum. The purpose of chromium is to increase the corrosion resistance. The role of molybdenum is grain refinement and strengthening of the matrix. In the substructure of Co-based alloys, a significant concentration of stacking faults is visible, which results from the low stacking fault energy value. Reducing the density of stacking faults can be achieved by the addition of tungsten [6]. Good wear resistance of cobalt alloys is usually attributed to increased strengthening and formation of carbides or nitrides. In biomedical applications, the carbon content has to be limited, to avoid the risk of the carbides' transformation into a brittle tetragonal intermetallic σ Co_7Cr_8 phase. Elements manufactured from Co-based alloys with low carbon guarantee safe use (there is no risk of unexpected cracking due to the fragility of the material), but at the same time they wear out faster.

In order to obtain Co-based alloys that are wear-resistant and non-fragile, new dynamically developing production techniques, for example, Selective Laser Sintering (SLS) [7–11], can be applied. SLS deviates from traditional methods of producing materials (e.g., foundry) due to the extreme temperature gradient. After the metal powder has been melted by the laser, the solidification is very fast. Therefore, it is reasonable to expect that alloys produced by SLS will have different properties and a micro/nanostructure compared to alloys that crystallize more slowly. In particular, cobalt-based alloys with homogeneous structure, good mechanical properties, and resistant to wear can be obtained. The second technique also offering such possibilities is Powder Injection Molding (PIM). PIM is a plastic shaping technique applied to powder metallurgy systems [12,13], where a small size powder is used that can be next sintered to full density. The aim of the present work was to analyze and compare the structure and wear resistance of Co-based alloy samples with low carbon content, produced by SLS and PIM.

2. Materials and Methods

The specimen produced by the SLS technique was obtained from metal powder (EOS CobaltChrome SP2) with the EOSINT M270 machine (Electro Optical Systems GmbH, Krailling, Germany), using the standard laser sintering parameters (laser power: 200 W; laser spot diameter: 200 μm; layer thickness: 20 μm; scan speed: up to 2.0 m/s). The chemical composition of the used powder is presented in Table 1. After additive manufacturing, the material was subjected to a thermal treatment typical of dental applications. Stress relieving thermal treatment was performed under argon atmosphere and consisted of heating to 450 °C in 60 min, maintaining this temperature for 45 min, heating to 750 °C in 45 min, and then maintaining this temperature for 60 min. Next, a fire simulation of the ceramic veneering, including five steps, was applied. In each step, the temperature was increased from 450 °C to about 950 °C for 5 minutes with vacuum. Subsequently, the material obtained was designated as Co-SLS. For the production of the PIM specimen, a CoCrMo powder with low carbon content (<0.05%) was used (Table 1). The powder was mixed with a two-component binder basis on paraffin and the skeleton polymer to produce the feedstock. The homogenized mixture was then injected by a piston injection molding machine into a mold and debinded by a two-step process (thermal degradation followed by solvent degradation in pure heptane at 20–40 °C). Thermal debinding was carried out in a flowing gas mixture Ar-10% H_2 at the rate of 0.1 L/min. The samples were sintered at the same atmosphere at a temperature of 1260 °C. A more detailed description of manufacturing is given in [14].

Table 1. Chemical composition of the powder (wt.%), used for Co-Selective Laser Sintering (Co-SLS) and Co-Powder Injection Molding (Co-PIM) manufacturing.

	Co	Cr	Mo	W	Si	Fe	Mn	C
SLS	63.8	24.7	5.1	5.4	1.0	<0.5	<0.1	<0.05
PIM	64.9	27.0	5.3	<0.2	<1.0	<0.75	<1	<0.05

Tribological tests were performed on a CSM tribometer (CSM Instruments, Peseux, Switzerland), using the "ball on disk" method based on the methodology presented in ASTM G99-95A [15] and ISO/TS 14569-2: 2001 [16]. During wear testing, both samples were rotated and kept in constant contact with the counter-sample in the form of a 6 mm diameter Al_2O_3 ball (Gewa, Zabrze, Poland). The experiment was carried out in distilled water at 37 ± 1 °C. The normal force applied by weight was 10 N, the relative sliding velocity was kept constant at 10 mm/s, the diameter of the track along which the sample was moving was 3 mm, and the total sliding distance was 100 m.

The X-ray diffraction (XRD) measurements were performed with the Bragg–Brentano camera geometry using an X'Pert Philips diffractometer with a generator voltage of 40 kV (Co-Kα radiation). The relative amounts of γ and ε phases were estimated by measuring the integrated intensities of the principal peaks (Equation (1)) [6]:

$$\% \ \gamma \ \text{phase} = \frac{I_{(200)\gamma}}{I_{(200)\gamma} + 0.45 I_{(101)\varepsilon}} \times 100 \tag{1}$$

where $I_{(200)\gamma}$ and $I_{(101)\varepsilon}$ are the diffracted intensities by (200)γ and (101)ε, respectively.

Microstructure and traces left after wear tests were investigated using a Zeiss Supra 35 field-emission Scanning Electron Microscope (SEM) (Carl Zeiss SMT, Oberkochen, Germany). For Transmission Electron Microscopy (TEM) observations, specimens were prepared by the conventional thinning procedure and by Focused Ion Beam (FIB) technique using the SEM/Ga-FIB Helios NanoLab™ 600i microscope (FEI, Hillsboro, OR, USA). Thin foils, after mechanical polishing, were finally thinned by an ion beam system (Gatan Inc., Pleasanton, CA, USA) using Ar ions at 3 kV. Final milling of lamellae was performed with the 1 keV ion Ga beam. TEM was carried out using a S/TEM TITAN 80-300 microscope equipped with a CEOS probe aberration corrector (FEI Company, Hillsboro, OR, USA). STEM images were collected with a 24.5-mrad probe semi-angle. Detector ranges were 25 mrad (Bright Field (BF)) and 47–200 mrad (High Angle Angular Dark Field (HAADF)). Crystal Maker and Single Crystal software (CrystalMaker Software Limited, version 10.4.1, Oxfordshire, UK) were used to simulate the crystal structure and diffraction patterns.

3. Results

3.1. Crystalline Structure and Microstructure of Co-SLS and Co-PIM

The phase composition of the Co-SLS and Co-PIM is shown in Figure 1. The result of the XRD analysis confirms for both materials the presence of the equilibrium phase ε and the metastable phase γ, but their shares are different. The γ phase has a face-centered cubic (fcc) lattice with a nominal parameter a = 0.3548 nm (COD ID: 96-900-8466, space group Fm-3m, space group number 225). The ε phase has a hexagonal close-packed (hcp) lattice with nominal parameters a = 0.25071 nm and c = 0.40686 nm (COD ID: 96-900-8492, space group P63/mmc, space group number 194) [17]. For both samples, there are no clear reflections from other phases, especially from carbides and the σ phase. Only an asymmetrical shape of peaks can indicate the presence of these precipitates. For the Co-SLS, the γ phase is dominant. Based on Equation (1), the amount of the γ phase can be estimated as 85.9%. For Co-PIM, the relative amount of γ is lower and is equal to 8.9%.

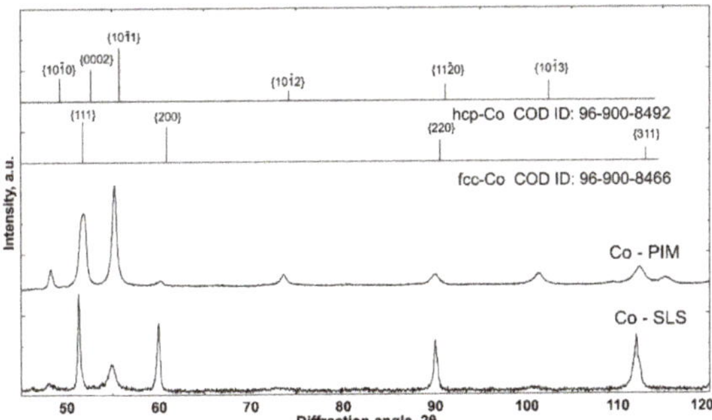

Figure 1. X-ray diffraction patterns of Co-Selective Laser Sintering (Co-SLM) and Co-Powder Injection Molding (Co-PIM) samples.

Microstructures obtained by SLS and PIM techniques are showed in Figure 2. The alloy obtained by SLS is homogeneous, without porosity, and with large (micrometric) precipitates. As a result of preparation for SEM observation (a mixture of nitric acid and hydrochloric acid was used), grain boundaries were revealed. The grains are irregular and vary in size and shape, but their size can be estimated at about 10–50 μm. The Co-PIM sample contains larger grains (approximately 20–60 μm). Uniformly distributed precipitates, especially on grain boundaries, are visible. However, it is worth emphasizing that the material obtained by PIM was also homogeneous and that no large pores were observed.

Figure 2. Scanning Electron Microscopy (SEM) images of the (**a**) Co-SLS and (**b**) Co-PIM samples.

The Co-SLS and Co-PIM structures were next analyzed by TEM. The substructure of Co-SLS was homogeneous, without visible pores and other large defects (Figure 3a), the presence of which was excluded earlier by SEM observations. Large precipitates of intermetallic phases (characteristic for cast alloy [18]) were still not detected here. The most characteristic feature of the Co-SLS structure is the presence of numerous stacking faults (SFs), arranged in two directions (indicated by red lines) and the presence of nanoprecipitates of two types (elongated and spherical), which are also evenly distributed (Figure 3b).

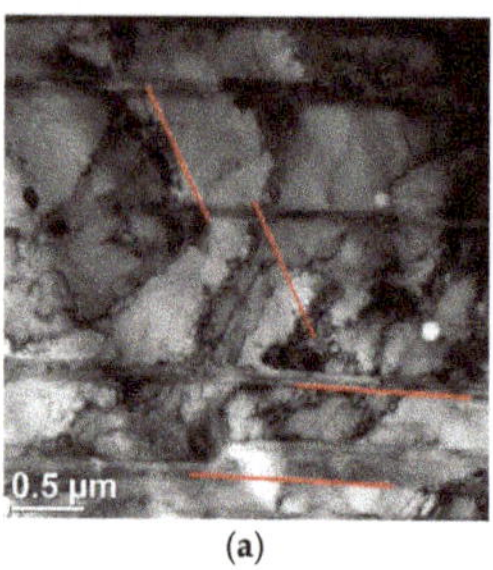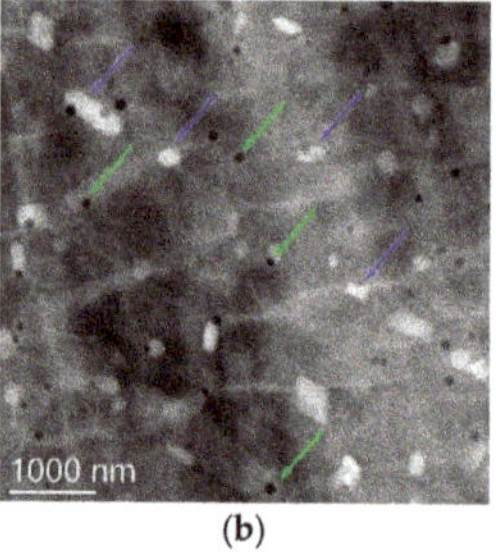

Figure 3. Co-SLS alloy sample structure. (**a**) Transmission Electron Microscopy Bright Field (TEM-BF) image, red lines indicate stacking faults (SF) and ε phase bands in the γ matrix; (**b**) Scanning Transmission Electron Microscopy High Angle Angular Dark Field (STEM-HAADF) image, blue arrows indicate nanoprecipitates of elongated shape, green arrows indicate spherical precipitates.

The Co-SLS matrix structure is shown in Figure 4, where intersecting ε phase plates with a width of several to several dozen nanometers (Figure 4a,c) and stacking faults (Figure 4b) are visible. For the [110] matrix orientation, the SFs intersect at 70.5° (Figure 4c). The phase composition of the matrix is also confirmed by the electron diffraction (Figure 4d). The results obtained show a pronounced tendency towards a deformation-induced phase transformation of the metastable γ phase to the hexagonal low-temperature ε phase, caused by a very low (and even negative) stacking fault energy of Co-based alloys [19]. A similar phenomenon—the TRansformation Induced Plasticity (TRIP) effect—has been described and is successfully used in steel design [20–22], enabling high strength properties while retaining high plasticity.

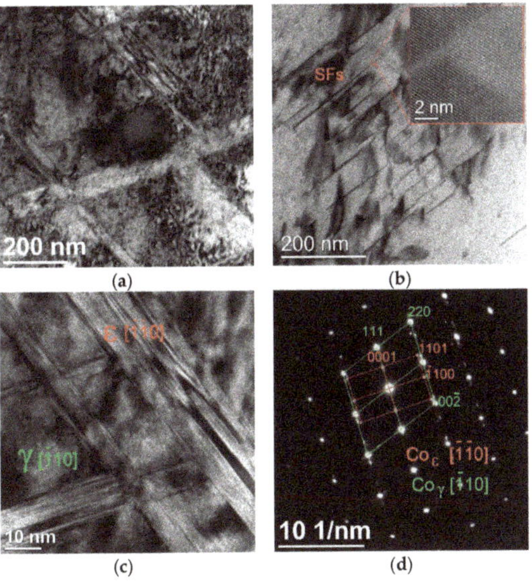

Figure 4. Co-SLS alloy matrix structure. (**a**) STEM-BF image; (**b**) slip bands, STEM-BF image; (**c**) ε phase bands in the γ matrix, TEM image; (**d**) Selected Area Electron Diffraction (SAED).

The irregular, elongated particles are evenly distributed, both at the grain boundaries and inside the grains (Figure 5a). Their size varies from 100 to 500 nm. Spot EDS analysis showed a higher W, Si, and Mo content in particles than in the matrix (Figure 5b, Table 2). A higher concentration of W, Mo, and Si was also confirmed by the distribution of these elements (Figure 5c).

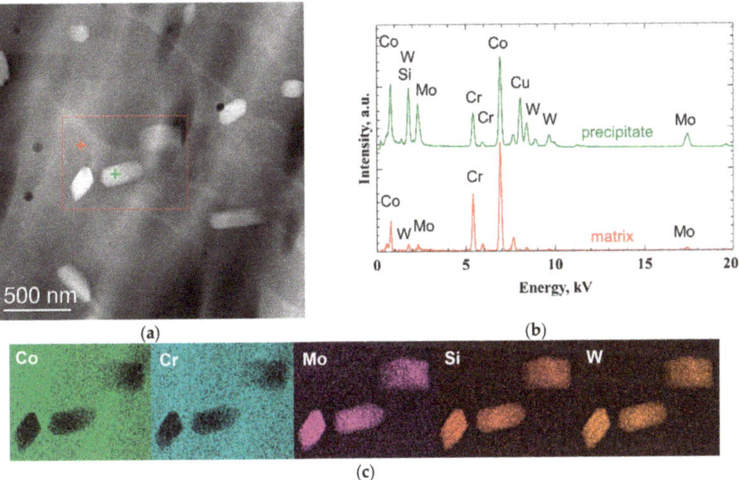

Figure 5. Chemical composition of the elongated shape precipitates. (**a**) STEM-HAADF image; (**b**) comparison of Energy Dispersive (EDS) spectra at the points indicated in (**a**); (**c**) maps of Co, Cr, Mo, Si, and W distribution in the area marked in (**a**).

Table 2. Experimental results (wt.%) of STEM-EDS analysis at the points indicated in Figure 4a.

Element	Co	Cr	Mo	Si	W
Matrix	65.9 ± 0.6	24.1 ± 0.4	4.6 ± 0.3	1.2 ± 0.1	4.2 ± 0.2
Elongated precipitate	41.1 ± 1.0	11.9 ± 0.5	21.5 ± 1.3	3.9 ± 0.6	21.6 ± 1.1
Spherical precipitate	62.2 ± 0.7	24.0 ± 0.4	4.1 ± 0.4	6.8 ± 0.3	2.9 ± 0.2

Earlier studies [23,24] indicated that the observed elongated particles have an hcp structure, with lattice parameters a = 0.4734 ± 0.0002 nm and c = 0.7661 ± 0.0003 nm and a composition resembling $Co_3(Mo,W)_2Si$, with possible interchange of Mo and W atoms in the crystallographic lattice (space group P63/mmc, space group number 194). Selected Area Electron Diffraction (SAED) obtained for three differently oriented precipitates (Figure 6a–c) confirm this statement. Co_3W_2Si is the $MgZn_2$-type Laves phase. Other examples of such structures are W_2Ni_3Si, Ti_2Ni_3Si, Mo_2Ni_3Si, and Co_3Mo_2Si [3,4,6,19,25,26]. These materials are known for their high hardness and dominant covalent nature of bonds, which makes them attractive in applications requiring high wear resistance [27,28]. The results of STEM imaging also showed that these hard precipitates in the Co-SLS can be deformed. Slip bands (marked with red arrows) are visible in Figure 7a. It confirms that the precipitates can, at least to some extent, change their shape before they break out. The second factor positively influencing the mechanical properties of the examined alloy is the presence of a cobalt matrix with the dominant share of the γ phase. Cobalt is soft, and there are even more slip systems in the γ phase than in the ε phase. Figure 7b confirms deformation of the matrix near the boundary of the hard precipitate.

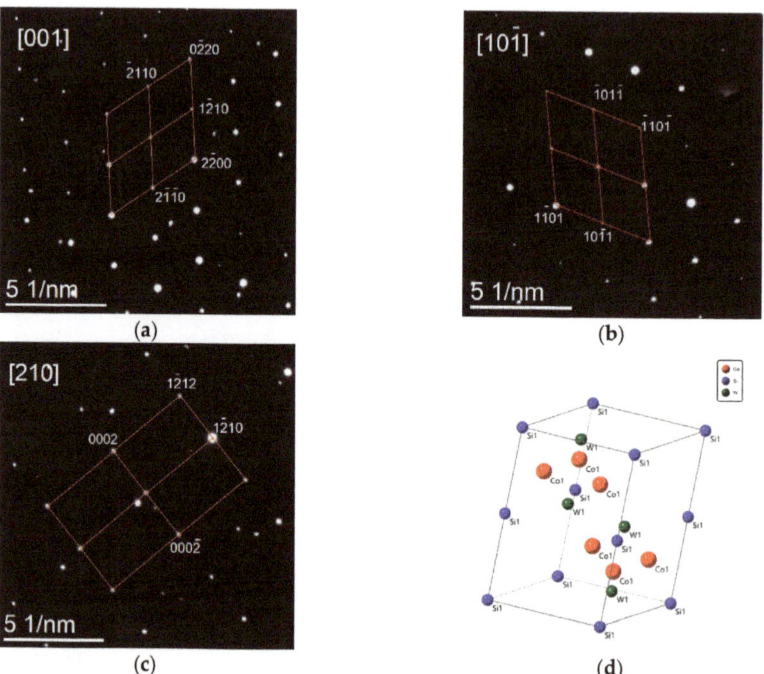

Figure 6. SAED electron diffraction of Co3(Mo,W)2Si in the direction (a) [001]; (b) [10−1]; and (c) [210]. (d) Model of Co3W2Si unit cell.

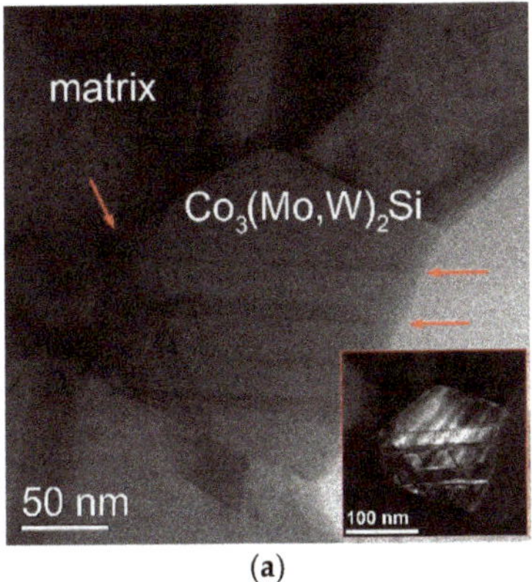
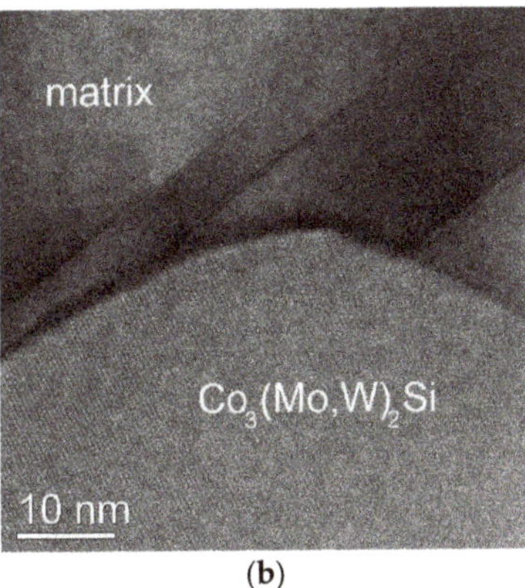

(**a**) (**b**)

Figure 7. (**a**) STEM-HAADF image of Co$_3$(Mo,W)$_2$Si precipitate with slip bands marked with arrows (the insert contains the DF image); (**b**) high-resolution (HR) STEM-HAADF image of the boundary between precipitate and matrix.

The observed precipitates of the second type are spherical. These precipitates are smaller, homogenous in size (50–100 nm), and evenly distributed. In HAADF images (Figures 3b and 5a) they are almost black, which could indicate empty spaces (pores). When using a larger magnification (Figure 8a), the presence of nanometric atomic clusters (with a diameter of approximately 2–5 nm) is observed. Chemical composition analysis, especially profile analysis (Figure 8b), confirms the increase (relative to the matrix) of Si concentration. It is worth emphasizing that this increase is statistically significant, but limited (in Figure 8b, to improve the readability of the figure, the intensity of the characteristic X-ray from Si has been increased six times). Figure 8c shows the HRTEM boundary between the spherical precipitate and the matrix. The figure also includes the Fast Fourier Transform (FFT) calculated for the matrix (upper) and precipitate (lower) part, respectively. Reflections are stronger on the first of them, which indicates a strongly crystalline matrix structure. In the case of the second FFT, individual reflections are visible and only in positions obtained for the matrix. The crystal structure is not visible in the precipitate. This can be interpreted that the analyzed precipitates have an amorphous structure (the weakened diffraction signal comes from the matrix material that is above/below the precipitate). The second, more convincing explanation is that these pores are a residue from silicon releases. Si was present in the CobaltChrome SP2 powder and evaporated during the remelting by laser. This is indicated by the spherical shape of the observed nanostructure and by the presence of small clusters of silicon atoms, visible in Figure 8a. Si precipitates are usually considered unfavorable in composites due to the risk of their uncontrolled growth. This does not apply to precipitates of such limited sizes (<100 nm) as in the described case, as well as to amorphous precipitates. An additional confirmation is the appearance of slip bands, visible in Figure 8d. They occur only in the matrix, and the presence of precipitates does not affect their behavior.

Four examples of precipitates in the Co-PIM sample are shown in Figure 9a. Based on the results of the EDS analysis (Figure 9b,c; Table 3), we can confirm the presence of both carbides and σ phase. Point A in Figure 9 indicates the carbide (high Cr content), B indicates the σ phase (Cr and Co in a similar amount), while C is the matrix (Co dominates). Figure 9 also shows the transformation of carbide into a brittle intermetallic σ (Co$_7$Cr$_8$) phase. The presence of the σ phase has been confirmed by SAED from the area marked in Figure 10. The σ phase has a tetragonal lattice with a nominal parameter a = 0.8810 nm and b = 0.4560 nm (COD ID: 98-010-2316, space group P42/mnm, space group number 136) [29].

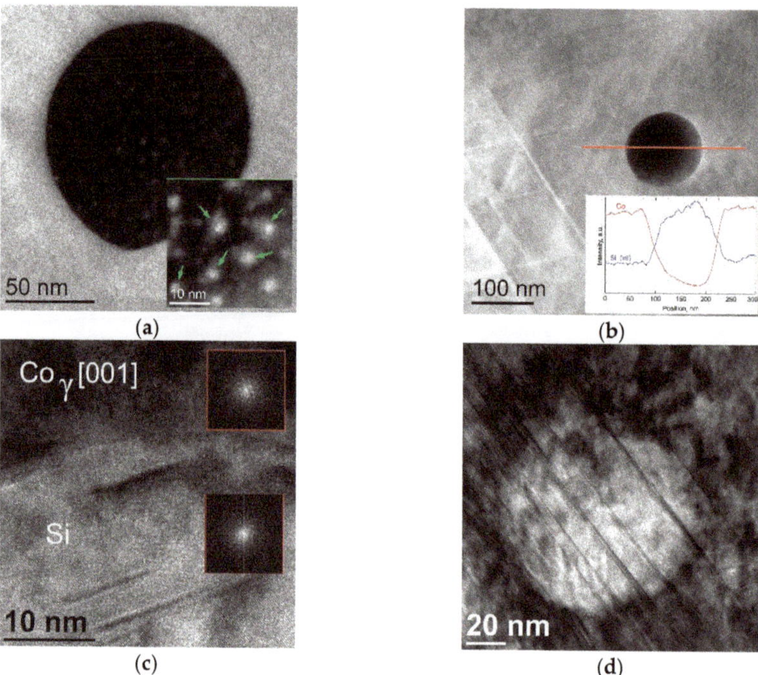

Figure 8. (**a**) STEM-HAADF image of Si-containing precipitation, the inset shows an enlargement with visible clusters of Si atoms, indicated by green arrows; (**b**) profile analysis of Co and Si exchanges; (**c**) STEM-BF image of the boundary between precipitate and matrix; (**d**) STEM-BF image of strongly deformed matrix with Si precipitate.

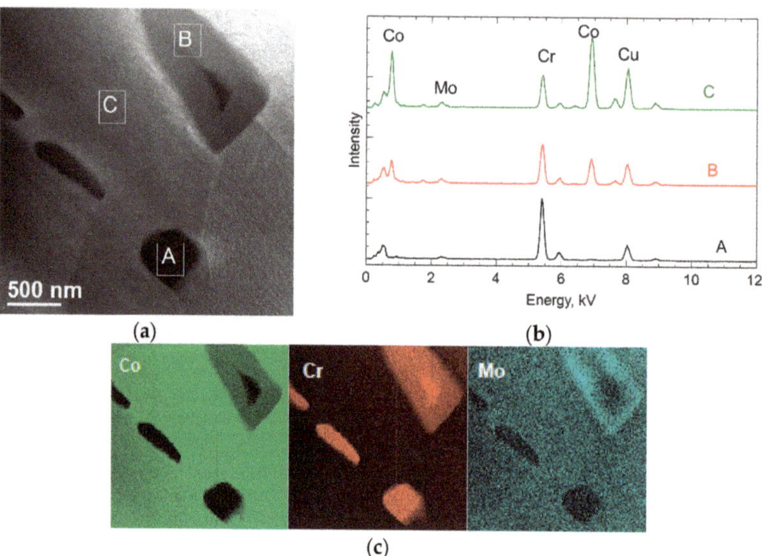

Figure 9. Chemical composition of the precipitates in the Co-PIM sample. (**a**) STEM-HAADF image; (**b**) comparison of EDS spectra at the points indicated in (**a**); (**c**) maps of Co, Cr, and Mo distribution in the area marked in (**a**).

Table 3. Experimental results (wt.%) of STEM-EDX analysis at the points indicated in Figure 9a. n.d. = not determined.

Element	Co	Cr	Mo	C	Si
A	2.0 ± 0.1	87.4 ± 1.0	10.6 ± 0.4	n. d.	0
B	40.4 ± 0.5	50.1 ± 0.5	7.9 ± 0.3	n. d.	1.6 ± 0.1
C	68.2 ± 0.5	26.6 ± 0.3	5.2 ± 0.2	n. d.	0

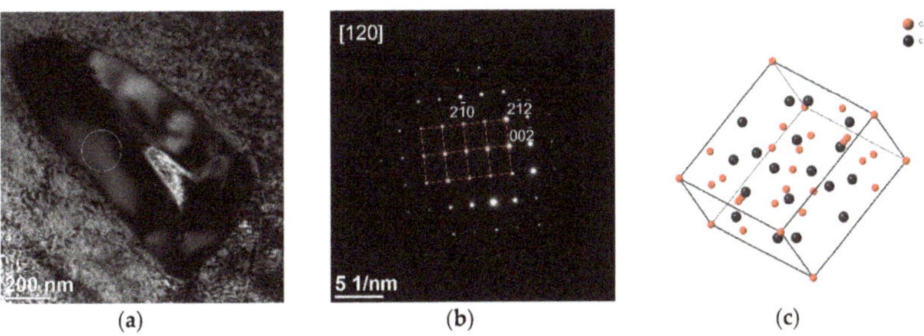

(a) (b) (c)

Figure 10. TEM image of selected precipitate in the Co-PIM sample (**a**). SAED electron diffraction of σ phase (Co$_7$Cr$_8$) in [120] the direction (**b**). Model of Co$_7$Cr$_8$ unit cell (**c**).

3.2. Wear Resistance of Co-SLS and Co-PIM

Figure 11 shows a comparison of friction coefficient changes during ball-on-disk wear testing for the Co-SLS and Co-PIM samples. The friction coefficient of the Co-SLS sample remained constant (approximately 0.4) throughout the measurement. In the first part of the test (up to 60 m), the friction coefficient of the PIM sample was about 0.6, and then increased to 0.8. The worn surfaces were then examined with SEM (Figure 12). In the case of the SLS sample, the cavities are small, which confirms the high wear resistance of the SLS alloy. In the Co-PIM sample, obvious abrasive wear is observed, which brought out a large volume of material from the surface, causing the formation of many pits on the surface.

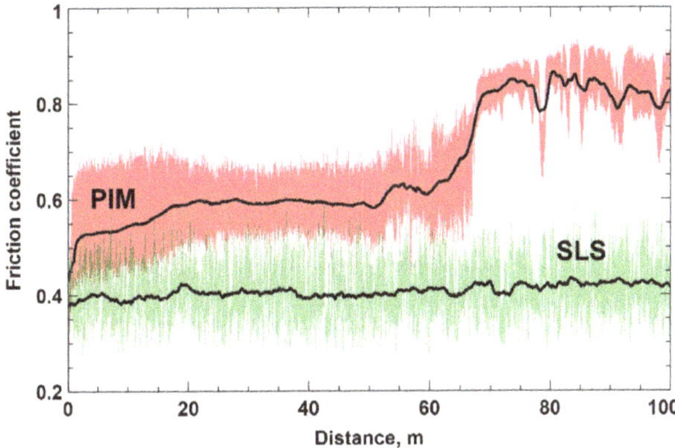

Figure 11. The friction coefficient during ball-on-disk wear testing.

 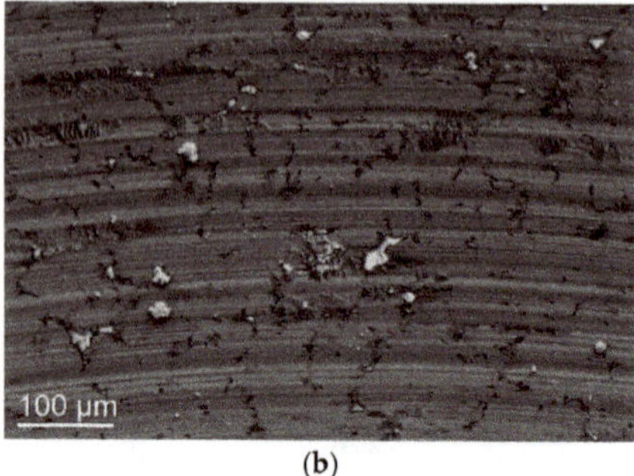

(a)   (b)

Figure 12. SEM images of wear scars of the (**a**) Co-SLS and (**b**) Co-PIM samples.

4. Discussion

Excellent wear resistance is one of the characteristics of Co-Cr alloys. Additive manufacturing for metals and powder injection molding have recently received much attention as techniques for producing complex three-dimensional structures with excellent wear resistance from precursor powders. Our research results make it possible to compare the structure and abrasive wear of materials made with these techniques.

Differences for materials obtained from powders using SLS and PIM were visible during wear resistance measurements. The friction coefficient of the Co-SLS sample had a lower value and remained constant throughout the measurement. For the PIM sample, the value of this coefficient was higher from the beginning and increased in the second half of the test. It is known that mechanical and tribological properties of Co-based alloys strongly depend on their microstructure [30–32]. For example, friction coefficient is mainly influenced by microstructure, and the roughness impact is less important [33]. The use of electron microscopy and XRD allowed us to show that Co-SLS and Co-PIM have different structures and to link these differences with observed wear behavior.

Both materials are homogeneous, compact, and without porosity. The grain size is limited in both cases. A smaller grain size leads to a higher yield strength for the alloy, according to the Hall–Petch equation. Smaller grain sizes indicate the presence of more grain boundaries per unit area. Such boundaries act as impeding points to dislocation propagation; as a result, greater applied stress is needed for the dislocation propagation. However, the main factor determining wear resistance of CoCrMo alloys is the carbon amount, the homogeneity of the carbide distribution, and the presence of the hcp crystal structure [34–36].

The phase composition and amount of the precipitates in the tested materials result from the chemical composition and from the manufacturing process. Powders used for both techniques included a low content of carbon (<0.05%). Most studies have suggested that a high carbon content in CoCrMo alloys improves the wear properties [30,37], resulting in excellent wear [38,39]. The presence of carbides affects the wear due to the hardness and the good coherency with the surrounding matrix, acting as a protective barrier against matrix delamination [30]. The small, evenly distributed carbides (especially on the surface) and the fine grain size have proven to be significantly beneficial for wear performance. The results of earlier studies do not indicate the improvement in wear behavior of Co-based alloys with low carbon content [38,39].

Better wear resistance of Co-SLS, however, does not result from the presence of carbides. Their role was taken over by hard and evenly distributed $Co_3(Mo,W)_2Si$ nanoprecipitates. The carbides were found in the Co-PIM sample, as a result of uncomplete polymer degradation. Their presence should be considered undesirable, primarily due to the limited control and the risk of transformation into a brittle σ phase.

A significant difference between the samples' structure concerns the phase composition of the matrix. As expected, the hcp ε phase is dominant in the matrix of the structure fabricated by PIM.

The fcc γ phase is the main matrix component of the SLS sample, what can be explained by the conditions of its production process. The critical value of thermal treatment for Co-based alloys, promoting martensitic transformation $\gamma \rightarrow \varepsilon$, is 970 °C. To a lesser extent, the rate of martensitic transformation depends on the time of thermal treatment, as demonstrated by Zhang [40]. Therefore, the high proportion of the γ phase in Co-SLS can be explained by the high cooling rate characteristic of SLS [1,2] and low temperature of heat treatment (750 °C). Alloys containing the ε phase (hcp) have lower wear friction and wear volume losses than alloys with a higher share of the γ (fcc) phase [1,31,32]. This effect is associated with few slip systems in the hcp structure and more difficult plastic deformation, which in turn avoids debris formation and results in better wear resistance. The results presented in [32] indicated that an alloy with an almost fully complete hcp structure exhibits a lower wear loss compared to an fcc structure. However, the high hardness of the carbides and/or other intermetallic precipitates could cause abrasive wear damage, when they are pulled out and fractured from the matrix [31]. Large precipitates are usually weakly bonded with the matrix. During wear resistance measurements they can be detached, with abrasive wear mechanism taking place. Large precipitates act as abrasives and reduce the alloy's wear resistance [32]. This was the case of Co-PIM. Abrasive wear was caused by precipitates such as the σ phase and/or carbides, contained in the ε matrix. For the Co-SLS sample, these negative factors were limited. Intermetallic precipitates strengthen the alloy, but to some extent they can be deformed, which reduces the probability of pulling them out of the matrix. An equally important factor is the possibility of matrix deformation near the precipitates. Because the precipitates are of limited size and are uniformly distributed in the matrix, the volume of torn-out material is also smaller. Moreover, the ε-martensite, formed by the strain-induced martensitic transformation through plastic straining during the sliding of the pin on the disk, decreases the wear rate [41].

5. Conclusions

New dynamically developing production techniques, for example, Selective Laser Sintering (SLS) and Powder Injection Molding (PIM), make it possible to produce homogeneous, compact, and non-porous Co-based alloys with low carbon content. An additional advantage of using these techniques is the ability to produce details of complex shapes. By combining the characteristic features of the matrix and the reinforcing phase, the SLS technique gains an additional advantage—high resistance to abrasive wear. To effectively use PIM for the production of Co-based alloy components, characterized by the required abrasive wear, the carbon content that remains as a result of incomplete polymer degradation must be controlled. Overcoming this barrier will expand the range of PIM applications.

Author Contributions: Conceptualization, A.Z., G.M. and M.P.; Investigation, K.M., W.P. and G.M.; Methodology, A.Z., K.M. and W.P.; Project administration, A.Z.; Writing—original draft, M.P.; Writing—review & editing, A.Z., K.M., G.M. and W.P. All authors have read and agreed to the published version of the manuscript.

Funding: This research was funded by the National Center of Science [Grant No. 2018/02/X/ST8/01839].

Conflicts of Interest: The authors declare no conflict of interest.

References

1. Saldívar García, A.D.J.; Medrano, A.M.; Salinas Rodríguez, A. Formation of hcp martensite during the isothermal aging of an fcc Co-27Cr-5Mo-0.05C orthopedic implant alloy. *Metall. Mater. Trans. A Phys. Metall. Mater. Sci.* **1999**, *30*, 1177–1184. [CrossRef]

2. Lopez, H.F. Alloy developments in biomedical co-base alloys for HIP implant applications. *Mater. Sci. Forum* **2013**, *736*, 133–146.

3. Yamanaka, K.; Mori, M.; Kurosu, S.; Matsumoto, H.; Chiba, A. Ultrafine grain refinement of biomedical Co-29Cr-6Mo alloy during conventional hot-compression deformation. *Metall. Mater. Trans. A Phys. Metall. Mater. Sci.* **2009**, *40*, 1980–1994. [CrossRef]

4. Talonen, J.; Hänninen, H. Formation of shear bands and strain-induced martensite during plastic deformation of metastable austenitic stainless steels. *Acta Mater.* **2007**, *55*, 6108–6118. [CrossRef]

5. Wei, W.; Zhou, Y.; Liu, W.; Li, N.; Yan, J.; Li, H. Microstructural Characterization, Mechanical Properties, and Corrosion Resistance of Dental Co-Cr-Mo-W Alloys Manufactured by Selective Laser Melting. *J. Mater. Eng. Perform.* **2018**, *27*, 5312–5320. [CrossRef]

6. Karaali, A.; Mirouh, K.; Hamamda, S.; Guiraldenq, P. Microstructural study of tungsten influence on Co-Cr alloys. *Mater. Sci. Eng. A* **2005**, *390*, 255–259. [CrossRef]

7. Marchese, G.; Garmendia Colera, X.; Calignano, F.; Lorusso, M.; Biamino, S.; Minetola, P.; Manfredi, D. Characterization and Comparison of Inconel 625 Processed by Selective Laser Melting and Laser Metal Deposition. *Adv. Eng. Mater.* **2017**, *19*, 1600635. [CrossRef]

8. Mindt, H.W.; Desmaison, O.; Megahed, M.; Peralta, A.; Neumann, J. Modeling of Powder Bed Manufacturing Defects. *J. Mater. Eng. Perform.* **2018**, *27*, 32–43. [CrossRef]

9. Kruth, J.P.; Mercelis, P.; Van Vaerenbergh, J.; Froyen, L.; Rombouts, M. Binding mechanisms in selective laser sintering and selective laser melting. *Rapid Prototyp. J.* **2005**, *11*, 26–36. [CrossRef]

10. Ziębowicz, A.; Woźniak, A.; Ziębowicz, B. The influence of technology on the physicochemical and electrochemical properties of the prosthetic materials. In Proceedings of the Advances in Intelligent Systems and Computing, Katowice, Poland, 18–20 October 2018; Springer: Cham, Switzerland; Volume 623, pp. 349–357.

11. Harris, I.D. A Transformational Advanced Manufacturing Technology. *Adv. Mater. Processes* **2012**, *25*.

12. German, R.M. Wear applications offer further growth for PIM. *Met. Powder Rep.* **1999**, *54*, 24–28. [CrossRef]

13. Wahi, A.; Muhamad, N.; Sulong, A.B.; Ahmad, R.N. Effect of Sintering Temperature on Density, Hardness and Strength of MIM Co30Cr6Mo Biomedical Alloy. *J. Jpn. Soc. Powder Powder Metall.* **2016**, *63*, 434–437. [CrossRef]

14. Matula, G.; Tomiczek, B.; Król, M.; Szatkowska, A.; Sotomayor, M.E. Application of thermal analysis in the selection of polymer components used as a binder for metal injection moulding of Co–Cr–Mo alloy powder. *J. Therm. Anal. Calorim.* **2018**, *134*, 391–399. [CrossRef]

15. *ASTM G99—95a Test Method for Wear Testing with a Pin-On-Disk Apparatus*; ASTM International: West Conshohocken, PA, USA, 2000.

16. ISO/TS 14569-2:2001. *Dental Materials—Guidance on Testing of Wear—Part 2: Wear by Two—And/or Three Body Contact*; ISO: Geneva, Switzerland, 2001.

17. Wyckoff, R.W.G. Cubic Closest Packed, ccp, Structure. In *Crystal Structures*, 2nd ed.; Interscience Publishers: New York, NY, USA, 1963; pp. 7–83.

18. Zhou, Y.; Wei, W.; Yan, J.; Liu, W.; Li, N.; Li, H.; Xu, S. Microstructures and metal-ceramic bond properties of Co-Cr biomedical alloys fabricated by selective laser melting and casting. *Mater. Sci. Eng. A* **2019**, *759*, 594–602. [CrossRef]

19. Weissensteiner, I.; Petersmann, M.; Antretter, T.; Clemens, H.; Weißensteiner, I.; Erdely, P.; Stark, A.; Maier-Kiener, V. Deformation-induced phase transformation in a Co-Cr-W-Mo alloy studied by high-energy X-ray diffraction during in-situ compression tests. *Acta Mater.* **2018**, *164*, 272–282. [CrossRef]

20. Frommeyer, G.; Métallurgie, O.G.-R. High strength TRIP/TWIP and superplastic steels: Development, properties, application. *Rev. Métallurgie* **1998**, *95*, 1299–1310. [CrossRef]

21. Zaefferer, S.; Ohlert, J.; Bleck, W. A study of microstructure, transformation mechanisms and correlation between microstructure and mechanical properties of a low alloyed TRIP steel. *Acta Mater.* **2004**, *52*, 2765–2778. [CrossRef]

22. Grässel, O.; Krüger, L.; Frommeyer, G.; Meyer, L.W. High strength Fe–Mn–(Al, Si) TRIP/TWIP steels development—Properties—Application. *Int. J. Plast.* **2000**, *16*, 1391–1409. [CrossRef]

23. Mengucci, P.; Barucca, G.; Gatto, A.; Bassoli, E.; Denti, L.; Fiori, F.; Czyrska-Filemonowicz, A. Effects of thermal treatments on microstructure and mechanical properties of a Co–Cr–Mo–W biomedical alloy produced by laser sintering. *J. Mech. Behav. Biomed. Mater.* **2016**, *60*, 106–117. [CrossRef]

24. Santecchia, E.; Gatto, A.; Bassoli, E.; Denti, L.; Rutkowski, B.; Mengucci, P.; Barucca, G. Precipitates formation and evolution in a Co-based alloy produced by powder bed fusion. *J. Alloys Compd.* **2019**, *797*, 652–658. [CrossRef]

25. Yamanaka, K.; Mori, M.; Koizumi, Y.; Chiba, A. Local strain evolution due to athermal $\gamma \rightarrow \varepsilon$ martensitic transformation in biomedical CoCrMo alloys. *J. Mech. Behav. Biomed. Mater.* **2014**, *32*, 52–61. [CrossRef] [PubMed]

26. Wang, H.; Luan, D.; Zhang, L.Y. Microstructure and wear resistance of laser melted W/W2Ni3Si metal silicides matrix in situ composites. *Scr. Mater.* **2003**, *48*, 1179–1184. [CrossRef]

27. Liu, Y.; Wang, H.W. Toughening and dry sliding wear resistance of Co3Mo2Si alloys. *Wear* **2013**, *297*, 952–957. [CrossRef]

28. Liu, Y.; Wang, H.M. Microstructure and wear property of laser-clad Co3Mo2Si/Coss wear resistant coatings. *Surf. Coat. Technol.* **2010**, *205*, 377–382. [CrossRef]

29. Dickins, G.J.; Douglas, A.M.B.; Taylor, W.H. The crystal structure of the Co–Cr σ phase. *Acta Crystallogr.* **1956**, *9*, 297–303. [CrossRef]

30. Varano, R.; Bobyn, J.D.; Medley, J.B.; Yue, S. The effect of microstructure on the wear of cobalt-based alloys used in metal-on-metal hip implants. *Proc. Inst. Mech. Eng. Part H J. Eng. Med.* **2006**, *220*, 145–159. [CrossRef]

31. Chiba, A.; Kumagai, K.; Nomura, N.; Miyakawa, S. Pin-on-disk wear behavior in a like-on-like configuration in a biological environment of high carbon cast and low carbon forged Co-29Cr-6Mo alloys. *Acta Mater.* **2007**, *55*, 1309–1318. [CrossRef]

32. Saldívar-García, A.J.; López, H.F. Microstructural effects on the wear resistance of wrought and as-cast Co-Cr-Mo-C implant alloys. *J. Biomed. Mater. Res. Part A* **2005**, *74*, 269–274. [CrossRef]

33. Bedolla-Gil, Y.; Hernandez-Rodriguez, M.A.L. Tribological behavior of a heat-treated cobalt-based alloy. *J. Mater. Eng. Perform.* **2013**, *22*, 541–547. [CrossRef]

34. Yan, Y.; Neville, A.; Dowson, D. Tribo-corrosion properties of cobalt-based medical implant alloys in simulated biological environments. *Wear* **2007**, *263*, 1105–1111. [CrossRef]

35. Long, W.T. The clinical performance of metal-on-metal as an articulation surface in total hip replacement. *Iowa Orthop. J.* **2005**, *25*, 10–16. [PubMed]

36. Cawley, J.; Metcalf, J.E.P.; Jones, A.H.; Band, T.J.; Skupien, D.S. A tribological study of cobalt chromium molybdenum alloys used in metal-on-metal resurfacing hip arthroplasty. *Wear* **2003**, *255*, 999–1006. [CrossRef]

37. Tipper, J.L.; Firkins, P.J.; Ingham, E.; Fisher, J.; Stone, M.H.; Farrar, R. Quantitative analysis of the wear and wear debris from low and high carbon content cobalt chrome alloys used in metal on metal total hip replacements. *J. Mater. Sci. Mater. Med.* **1999**, *10*, 353–362. [CrossRef] [PubMed]

38. Dowson, D.; Hardaker, C.; Flett, M.; Isaac, G.H. A hip joint simulator study of the performance of metal-on-metal joints: Part I: The role of materials. *J. Arthroplast.* **2004**, *19*, 118–123. [CrossRef]

39. Scholes, S.C.; Unsworth, A. Pin-on-plate studies on the effect of rotation on the wear of metal-on-metal samples. *J. Mater. Sci. Mater. Med.* **2001**, *12*, 299–303. [CrossRef]

40. Zhang, M.; Yang, Y.; Song, C.; Bai, Y.; Xiao, Z. An investigation into the aging behavior of CoCrMo alloys fabricated by selective laser melting. *J. Alloys Compd.* **2018**, *750*, 878–886. [CrossRef]

41. Hassani, F.Z.; Ketabchi, M.; Bruschi, S.; Ghiotti, A. Effects of carbide precipitation on the microstructural and tribological properties of Co–Cr–Mo–C medical implants after thermal treatment. *J. Mater. Sci.* **2016**, *51*, 4495–4508. [CrossRef]

Article

The Effect of Lath Martensite Microstructures on the Strength of Medium-Carbon Low-Alloy Steel

Chen Sun [1,2], Paixian Fu [1,*], Hongwei Liu [1], Hanghang Liu [1], Ningyu Du [1,2] and Yanfei Cao [1,*]

[1] Shenyang National Laboratory for Materials Science, Institute of Metal Research, Chinese Academy of Sciences, Shenyang 110016, China; csun15s@imr.ac.cn (C.S.); hwliu@imr.ac.cn (H.L.); hhliu15b@imr.ac.cn (H.L.); nydu16s@imr.ac.cn (N.D.)

[2] School of Materials Science and Engineering, University of Science and Technology of China, Hefei 230026, China

* Correspondence: pxfu@imr.ac.cn (P.F.); yfcao10s@imr.ac.cn (Y.C.)

Received: 7 March 2020; Accepted: 19 March 2020; Published: 23 March 2020

Abstract: Different austenitizing temperatures were used to obtain medium-carbon low-alloy (MCLA) martensitic steels with different lath martensite microstructures. The hierarchical microstructures of lath martensite were investigated by optical microscopy (OM), electron backscattering diffraction (EBSD), and transmission electron microscopy (TEM). The results show that with increasing the austenitizing temperature, the prior austenite grain size and block size increased, while the lath width decreased. Further, the yield strength and tensile strength increased due to the enhancement of the grain boundary strengthening. The fitting results reveal that only the relationship between lath width and strength followed the Hall–Petch formula of. Hence, we propose that lath width acts as the effective grain size (EGS) of strength in MCLA steel. In addition, the carbon content had a significant effect on the EGS of martensitic strength. In steels with lower carbon content, block size acted as the EGS, while, in steels with higher carbon content, the EGS changed to lath width. The effect of the Cottrell atmosphere around boundaries may be responsible for this change.

Keywords: medium-carbon low-alloy steel; lath martensite; effective grain size; strength; carbon content

---

1. Introduction

Medium-carbon low-alloy (MCLA) steel has an excellent combination of strength, toughness and hardenability and is widely used in structural components with large sections, such as generator spindles and automotive crankshafts [1–3]. Lath martensite is a typical microstructure seen in MCLA steel after quenching. There are several elements in lath martensitic microstructures. Prior austenite grains (PAGs) are divided into several packets, which consist of parallel blocks. The blocks are composed of laths, arranged parallel to each other. Low-angle grain boundaries (LAGBs) exist among laths, while high-angle grain boundaries (HAGBs) exist among packets and blocks [4–6].

An early work [7] indicated that the PAG size was the effective grain size (EGS) of strength in the Hall–Petch formula. Subsequently, the Hall–Petch relationship between yield strength and packet size was discovered by Swarr et al. in Fe–0.2C alloy [8], by Roberts in Fe–Mn alloy [9], and by Wang et al. in 17CrNiMo6 steel [10]. With the development of characterization technology, Morito et al. [11] revealed that the block width in Fe–0.2C and the Fe–0.2C–2Mn alloys is the key structural parameter controlling the strength of lath martensite by utilizing electron backscattered diffraction (EBSD). Similar results were presented by Zhang et al. [12], Long et al. [13], and Li et al. [14]. However, the martensite lath was also considered to be an effective control unit of strength in the research of Smith et al. in 42CrMo steel [15] and Kim et al. in Fe-0.55C alloy [16]. So far, there are still controversies regarding the EGS of strength in lath martensite.

Previous studies focused mainly on low-carbon steels. However, reports are scarce on the relationship between lath martensite microstructures and strength in medium-carbon steels has been. In lath martensite, carbon atoms segregate around dislocations and grain boundaries, rather than interstitial solution in the lattice [17]. The different carbon contents in steel can change the amount of segregated carbon atoms around the grain boundaries, which therefore affects its strength. In fact, research on low-carbon steel [8–14] draws the conclusion that the block/packet size is the key structural parameter in controlling the strength of lath martensite, while research on medium-carbon steel [16,18] tends to suggest that the strength depends primarily on the lath width. However, little work has been done on medium-carbon steel. Therefore, it is necessary to study the relationship between the martensite multi-level microstructure and strength in medium-carbon steel in order to verify the above finding.

In this work, austenitizing temperatures of 850–940 °C were used to obtain different sizes of PAG, block, and lath in the experimental MCLA steel. Then, optical microscopy (OM), EBSD, and transmission electron microscopy (TEM) were utilized to quantify the multi-level structural parameters at different austenitizing temperatures. The classical Hall–Petch formula of strength was assessed with PAG size, block width, and lath width respectively in order to clarify the EGS that governs the strength in the experimental MCLA steel. In addition, the influence of carbon content on the effective grain size of strength is summarized and further elucidated based on the above results as well as data from published research.

2. Experimental Procedures

2.1. Materials and Heat Treatment

The MCLA martensite steel used in this investigation was melted in a vacuum furnace and cast into a 25 kg ingot. Then, the ingot was forged into a round rod. Specimens with a dimension of 60 mm × 60 mm × 60 mm were taken from the rods. The chemical composition of the experimental steel was determined by an inductively coupled plasma emission spectrometer (ICP-6300, Thermo Fisher Scientific Company, Waltham, USA), and the result is shown in Table 1. Figure 1 shows the dilatometric curve of test steel, indicating that the Ac_1 (austenitization starting point) and Ac_3 (austenitization ending point) were respectively 735 °C and 818 °C. Based on the dilatometric curve, the heat treatment processes were as follows. First, the specimens were annealed at 860 °C for 3 h, followed by furnace cooling. Then the specimens were austenitized at different temperatures of 850, 880, 910, and 940 °C for 3 h, and quenched by water cooling.

2.2. Microstructure Observation

The PAGs were observed via OM (AxioCam MRc5, ZEISS Company, Oberkochen, Germany). The OM specimens were etched with a supersaturated picric acid solution, which was configured with 25 g of water and 0.7 g of picric acid. The martensite packets and blocks were characterized with EBSD. As the lath width is generally 0.2–0.3 µm [5,12,13], the step size was chosen to be 0.2 µm. The block widths were measured with reference to EBSD maps, as processed with the Oxford Instruments Channel 5 HKL software. The martensite laths were observed via TEM (Tecnai G220, FEI Company, Hillsboro, AL, USA). At least 500 PAGs, 300 blocks, and 200 laths were measured in order to ascertain the average PAG size, block width, and lath width. The preparation methods used for the TEM samples and EBSD samples can be found in our previous research [3]. The dislocation densities of experimental steels were determined via X-ray diffraction (XRD) (D/Max-2500PC, Rigaku Company, Tokyo, Japan) with Cu Kα radiation (λ = 1.5406 Å). The scanning angle was 20–100°, and the scanning speed was 1°/min.

Table 1. Chemical compositions of the experimental medium-carbon low-alloy (MCLA) martensite steel (wt%).

C	Si	Mn	Cr	Mo	S	P	Ni	V
0.41	0.26	0.70	1.10	0.26	0.0003	0.010	0.55	0.19

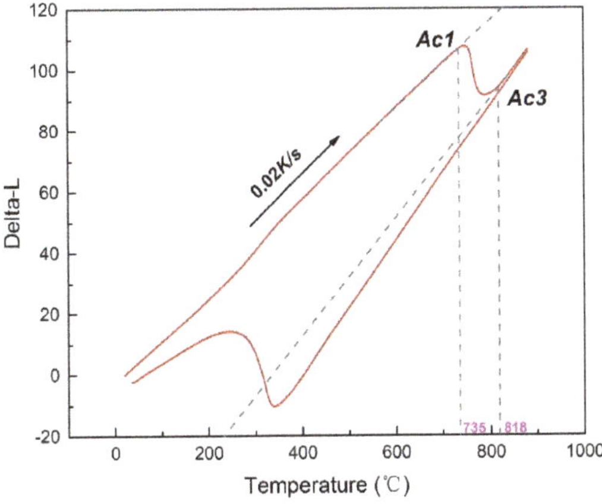

Figure 1. The dilatometric curve of the experimental MCLA steel.

2.3. Mechanical Tests

The tensile tests were performed with a Z150 tensile machine (Zwick/Roell Company, Ulm, Germany). At least three standard tensile samples with a diameter of 5 mm were used for each heat treatment condition.

3. Results and Discussion

3.1. Microstructural Characterization

Figure 2 shows the typical morphologies of PAG in MCLA martensitic steel when austenitized at different temperatures. The average PAG sizes of samples austenitized at different temperatures are shown in Figure 3a. The PAGs were uniform after austenitizing at 850 °C, as shown in Figure 2a. As the austenitizing temperature rises to 880 °C some PAGs coarsened severely, leading to an increase in the average PAG size. When the austenitizing temperature was increased further, the dissolution of fine precipitates weakened the pinning effect on the grain boundary [13], resulting in the fast growth rate of PAG size, as shown in Figure 3a.

The EBSD orientation maps of MCLA steel austenitized at different temperatures are shown in Figure 4, and the measured block width (d_B) is shown in Figure 3b. It can be seen that the PAGs are composed of several martensite packets which consist of blocks with similar extension directions. The PAG boundaries are HAGBs with a misorientation less than 45° (black lines) [13]. The HAGBs with a misorientation higher than 45° (yellow lines) are the martensite packet and block boundaries, which are distributed inside the PAGs. As the austenitizing temperature increased, the martensite blocks were arranged in a more orderly fashion. This is due to the weakening of the resistance of the lath nucleation and growth at high austenitizing temperatures [13]. At the same time, the sizes of the packets and blocks increased significantly. As shown in Figure 3b, as the austenitizing temperature increased from 850 °C to 940 °C, the average d_B increased from ~1.8 μm to ~2.5 μm.

The TEM observations of martensite laths in as-quenched MCLA steels are shown in Figure 5, and the measured lath width (d_L) is shown in Figure 3c. The laths arrange in a parallel manner in packets, and contain high-density dislocations. The d_L gradually decreased with increasing austenitizing temperature. The size distribution of d_L is shown in Figure 6. It can be seen that all distribution curves tend to show a normal distribution. The peak of the normal distribution curve moved to the left as the austenitizing temperature increased, revealing that higher austenitizing temperature resulted in a fine lath width. The high austenitizing temperature promotes the dissolution of residual carbides into austenite, which decreases the martensite starting temperature and increases the nucleation rate of martensite [19,20]. The low martensite starting temperature, coupled with the high nucleation rate, result in small d_L [13,21].

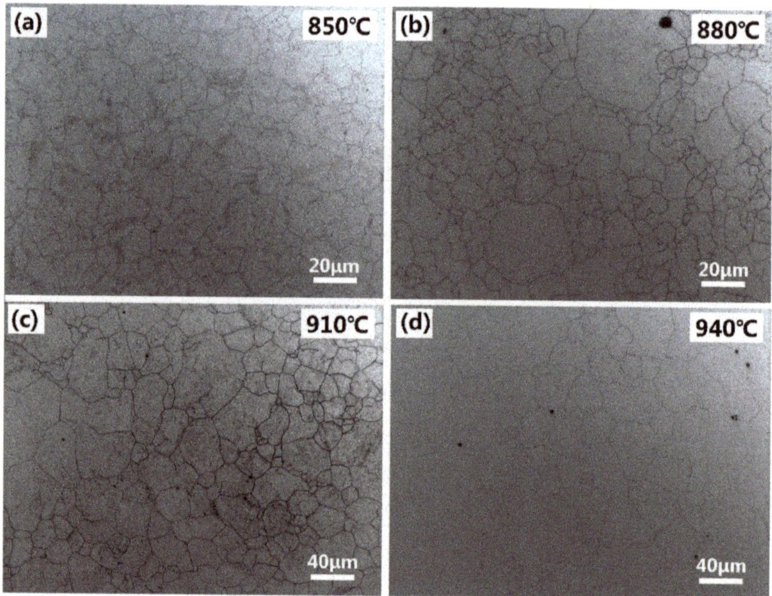

Figure 2. Prior austenite grain of MCLA martensitic steel austenitized at the different temperatures of (**a**) 850 °C, (**b**) 880 °C, (**c**) 910 °C, and (**d**) 940 °C observed by optical microscopy.

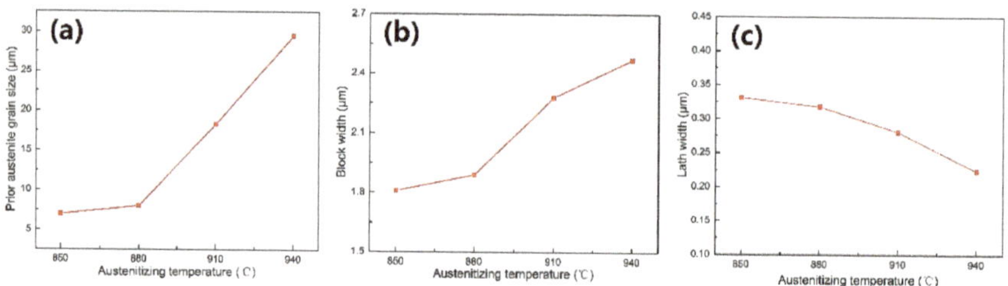

Figure 3. Relationship between austenitizing temperature and (**a**) prior austenite grain size, (**b**) block width, and (**c**) lath width.

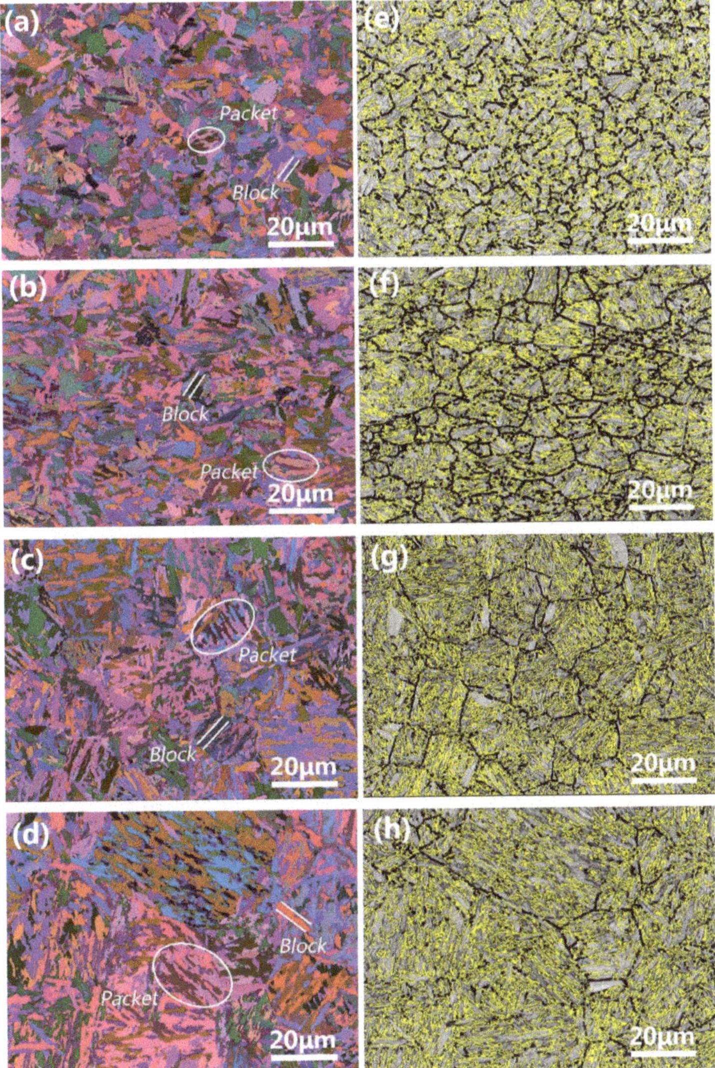

Figure 4. Electron backscattered diffraction (EBSD) of all Euler maps and the corresponding band contrast maps of MCLA martensitic steel austenitized at different temperatures of (**a**,**e**) 850 °C; (**b**,**f**) 880 °C; (**c**,**g**) 910 °C; (**d**,**h**) 940 °C (black lines: $45° > \theta > 15°$; and yellow lines: $\theta > 45°$).

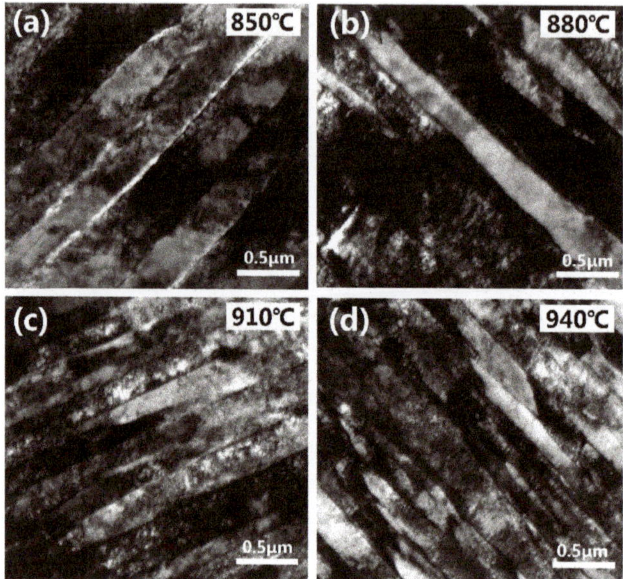

Figure 5. Transmission electron microscopy (TEM) micrographs of MCLA steel austenitized at the different temperatures: (**a**) 850 °C, (**b**) 880 °C, (**c**) 910 °C, and (**d**) 940 °C.

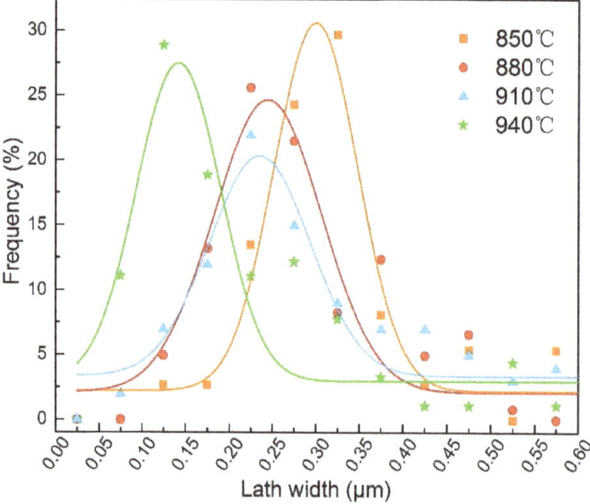

Figure 6. Size distribution of lath width at different austenitizing temperatures.

Figure 7 shows the relationship between d_B, d_L, and PAG size, respectively. The d_B increased linearly and the d_L decreased linearly with the PAG size at different austenitizing temperatures. For example, the d_B changed from 2.28 μm to 2.47 μm as the PAG size increased from 18.3 μm to 29.5 μm, while the d_L decreased from 281 nm to 224 nm. Similar results were reported by Long et al. [13].

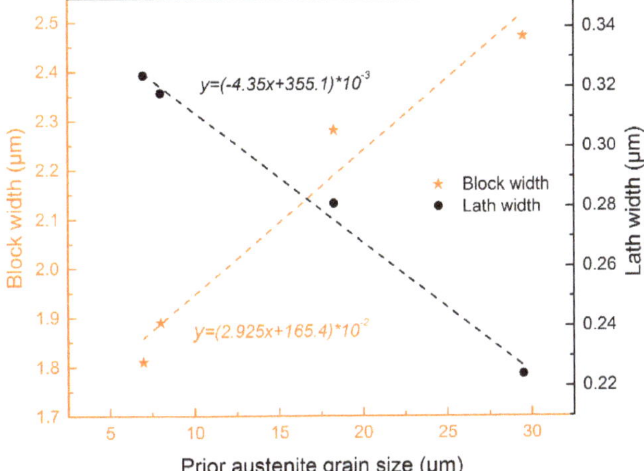

Figure 7. Dependence of block width and lath width on prior austenite grain size in MCLA steel.

3.2. Tensile Properties

Figure 8 shows the relationship between the tensile properties and the austenitizing temperature. When the quenching temperature was increased from 850 °C to 940 °C, the yield strength increased from 1510 MPa to 1591 MPa while the tensile strength increased from 2120 MPa to 2244 MPa. The elongation and section shrinkage were similar for all specimens, with the former being ~8% and the latter being ~35%. The high density of dislocations and boundaries restricted the movement of dislocations, resulting in the poor ductility of all of the samples.

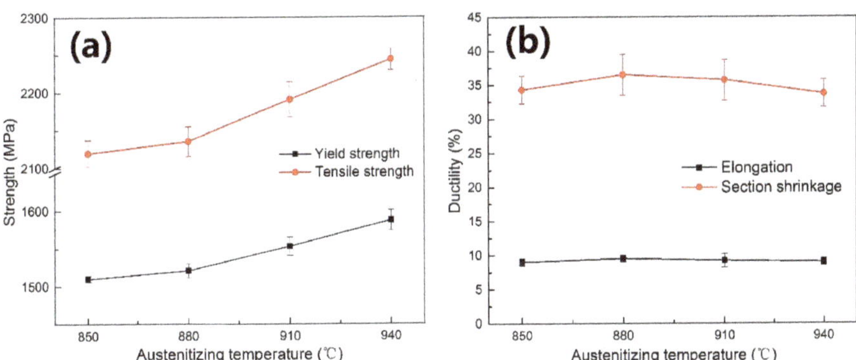

Figure 8. Tensile properties of MCLA martensite steel at different austenitizing temperatures: (**a**) yield strength and tensile strength; (**b**) elongation and section shrinkage.

3.3. Effect of Lath Martensite Microstructures on the Strength

Four strengthening contributions were considered in this work: solid-solution strengthening ($\Delta\sigma_s$), precipitation strengthening ($\Delta\sigma_p$), dislocation strengthening ($\Delta\sigma_d$), and grain boundary strengthening ($\Delta\sigma_g$). The yield strength can be expressed as Equation (1) [22,23]:

$$\sigma_{YS} = \Delta\sigma_0 + \Delta\sigma_s + \Delta\sigma_p + \Delta\sigma_d + \Delta\sigma_g \tag{1}$$

where σ_{YS} is the yield strength (MPa) and $\Delta\sigma_0$ is the intrinsic strength of the matrix (MPa), which was estimated to be 85~88 MPa [24,25].

The precipitation strengthening can be expressed as Equation (2) in light of the Orowan relationship [26,27]:

$$\Delta\sigma_p = \frac{0.538Gbf^{0.5}}{D} ln\left(\frac{D}{2b}\right) \tag{2}$$

where f is the volume fraction of the carbide, D is the mean particle size, and b is the Burgers vector of 0.248 nm. In this study, the carbides dissolved almost completely into the austenite and almost no precipitates remained in the microstructure. Therefore, the value of $\Delta\sigma_p$ can be regarded as 0 MPa while $\Delta\sigma_s$ was similar in all specimens.

$\Delta\sigma_d$ can be expressed as Equation (3) [27]:

$$\Delta\sigma_d = \alpha MGb\rho^{0.5} \tag{3}$$

where M is the Taylor factor, α is a constant of 0.435, G is the shear modulus, and ρ is the dislocation density.

Figure 9 shows the XRD patterns of specimens austenitized at different temperatures. No diffraction peak of retained austenite could be observed in the XRD pattern, meaning that the of retained austenite content was very small. The dislocation densities were calculated according to the XRD results. The calculation method is drawn from references [28,29]. The measured dislocation densities are shown in Figure 9b. It is shown that the dislocation density only changed slightly with the increase of austenitizing temperature, which is consistent with reference [30]. According to Equation (3), the $\Delta\sigma_d$ values were considered to be identical. Therefore, the change of grain boundary strengthening led to an increase of yield strength when the austenitizing temperature was increased. Under such circumstances, the yield strength is used as the value of grain boundary strengthening.

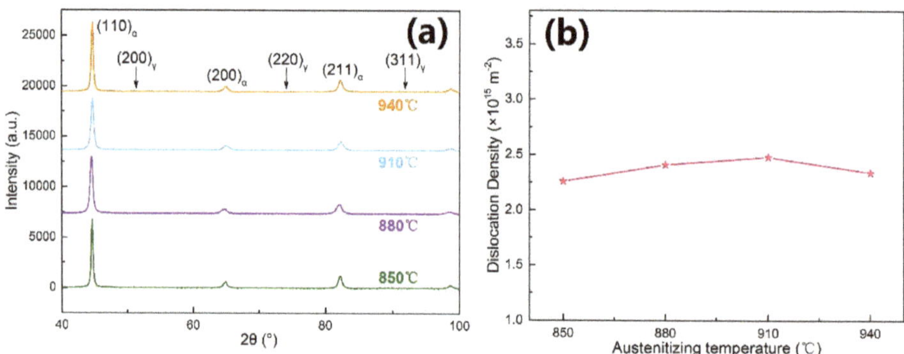

Figure 9. (a) X-ray diffraction (XRD) spectra and (b) dislocation density of MCLA martensitic steel austenitized at 850–940 °C.

The grain boundary strengthening can be described by the classic Hall–Petch relationship:

$$\sigma_{Ys} = \sigma_0 + k_y d^{-0.5} \tag{4}$$

where k_y is the Hall–Petch slope and d is the EGS. Evidently, the smaller the EGS, the more the boundaries can hinder dislocation motion, and the higher the strength. The relationship between the strength and lath martensite microstructure sizes is shown in Figure 10. The strength decreased linearly with $d_R^{-0.5}$ and $d_B^{-0.5}$, while it increased linearly with $d_L^{-0.5}$. That is, only the relationship between strength and lath width followed the Hall–Petch formula. In other words, the lath width was finally determined as the EGS of strength in the experimental MCLA steel.

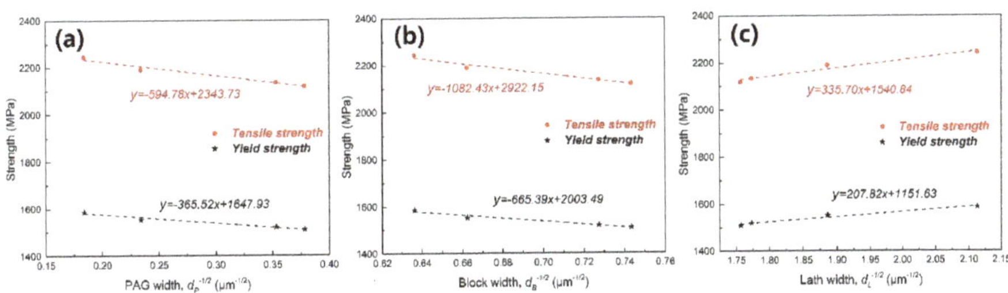

Figure 10. Strength as a function of (**a**) prior austenite grain (PAG) size (d_R), (**b**) block width (d_B), and (**c**) lath width (d_L).

In lath martensite, the PAG boundaries, packet boundaries, and block boundaries are HAGBs, while the lath boundaries are LAGBs. Lath width acting as the EGS means that the LAGBs play a dominant role in hindering dislocation motion. Many previous studies have suggested that only HAGBs can hinder the dislocation motion effectively and cause strengthening [31–33]. However, recent research has come to the opposite conclusion. Du et al. [34] demonstrated that both high- and low-angle grain boundaries act as effective barriers to dislocation movement via uniaxial micro-tensile tests. Chen et al. [35] directly observed that a large number of dislocations accumulate in front of the lath boundaries via in situ TEM experiments, proving that the LAGB is capable of hindering dislocation motion and causing strengthening. These findings act as further proof of our results showing that lath width can act as the EGS in MCLA steel.

3.4. Effect of Carbon Content on the Effective Grain Size of Strength

Figure 11 summarizes the relationship between strength and lath martensite microstructure sizes in recent studies [13,14,16,18,30]. As shown in Figure 11a, with increasing of $d_B^{-0.5}$, the strength increased in steels with carbon content below 0.2 wt%, while it decreased linearly in steels with a carbon content above 0.4 wt%. Figure 11b shows the relationship between strength and lath width, revealing the opposite result to that shown in Figure 11a. The relationship between the lath width and strength only followed the Hall–Petch relationship in the medium-carbon steels. Thus, it can be concluded that the EGS of strength seems to be related to the carbon content. In steels with lower carbon content the block size acted as the EGS. While, in steels with higher carbon content, the EGS changed to lath width. That is, the carbon content of steel affects the role of HAGB and LAGB in preventing dislocation movement. When the carbon content is low, HAGBs are the most significant barriers to the dislocation motion, and consequently the hindering effect of LAGB can be ignored. With an increase in carbon content, the hindering effect of LAGB on dislocation movement becomes more important, leading to the gradual transformation of EGS to lath width.

The effect of the Cottrell atmosphere around boundaries may be responsible for this change. Carbon atoms segregate around boundaries and dislocations to reduce distortion energy, leading to the formation of the Cottrell atmosphere [36–38]. During the slipping process, the dislocations are forced to move along with the Cottrell atmospheres, resulting in the so-called "drag effect", which can effectively increase the resistance of the dislocation movement and pin the dislocation. Previous studies have revealed that the drag effect enhances when the carbon content is increased [36,39]. When the carbon content is low (lower than 0.2 wt%), the drag effect of the Cottrell atmosphere becomes weak and the hindering effect of grain boundaries on dislocation motion is mainly dependent on its distorted lattice structure. In this case, the barrier effect of HAGB on dislocation movement is much stronger than that of LAGB. Accordingly, the block width acts as EGS in steels with low carbon content. With the increase of carbon content in steel, the drag effect of the Cottrell atmosphere is greatly enhanced. Accordingly, the difference between the hindering effect of HAGB and LAGB on the dislocation motion, which is caused by the lattice structure, is reduced. When the carbon content is high enough (higher than 0.4 wt%), the LAGB, which holds the absolute advantage in quantity, becomes the dominant factor in hindering dislocation movement. Therefore, the lath width becomes the EGS of strength in steels with high carbon content.

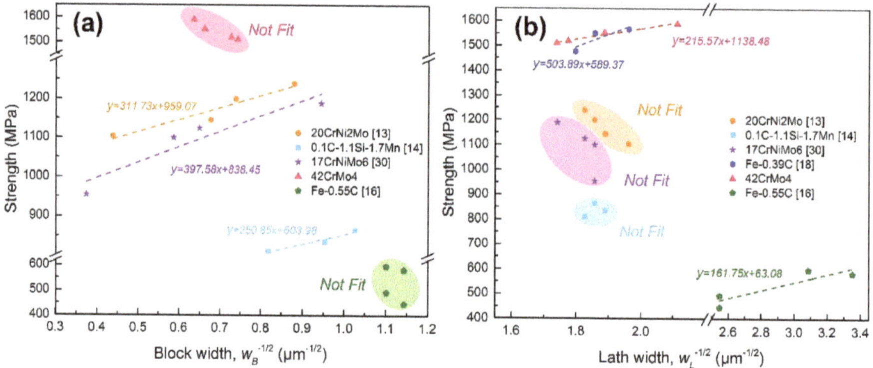

Figure 11. Yield strength as a function of (**a**) block width and (**b**) lath width.

4. Conclusions

We studied the effect of martensite microstructures on the strength of MCLA steel. The results are summarized as follows:

1. When increasing the austenitizing temperature, the PAG size and block size become larger, while the lath width decreases in the experimental MCLA steel. Both the increment of block width and the reduction of the lath width show a linear variation with the increase of the PAG size.
2. When the austenitizing temperature rises, the yield strength and the tensile strength are elevated due to the enhancement of the grain boundary strengthening. The Hall–Petch fitting results reveal that only the relationship between lath width and strength follows the Hall–Petch formula, which indicates that the lath width is the effective grain size of strength in the experimental MCLA steel.
3. Carbon content has a significant effect on the EGS of strength in lath martensite. In terms of low-carbon steels with a carbon content lower than 0.2 wt%, block size acts as the effective grain size, while the EGS tends to be lath width in steels with a high carbon content of over 0.4 wt%. The effect of the Cottrell atmosphere around boundaries is considered to be responsible for this change.

Author Contributions: Conceptualization, C.S., Y.C., and P.F.; methodology, C.S.; validation, C.S., Y.C., and P.F.; formal analysis, C.S.; investigation, C.S., Y.C., and P.F.; resources, P.F. and H.L. (Hongwei Liu); data curation, C.S. and P.F.; writing—original draft preparation, C.S.; writing—review and editing, C.S., Y.C., H.L. (Hanghang Liu), H.L. (Hongwei Liu), and N.D.; visualization, C.S., N.D. and H.L. (Hanghang Liu); project administration, P.F. and H.L. (Hongwei Liu); funding acquisition, H.L. (Hongwei Liu), P.F., and H.L. (Hanghang Liu) All authors have read and agreed to the published version of the manuscript.

Funding: This research was funded by the National Natural Science Foundation of China (Grant No. 51701225), the Project to Strengthen Industrial Development at the Grass-roots Level (Grant No.TC190A4DA/35), the Independent & Controllable Manufacturing of Advanced Bearing, the Innovation project of the cutting-edge basic research and key technology by SYNL (Grant No.L2019R36), the Young Talent Project by SYNL (Grant No.L2019F33) and the China Postdoctoral Science Foundation (Grant No. 2019M661153).

Conflicts of Interest: The authors declare no conflicts of interest.

References

1. Morris, J.W., Jr.; Kinney, C.; Pytlewski, K.; Adachi, Y. Microstructure and cleavage in lath martensitic steels. *Sci. Technol. Adv. Mater.* **2013**, *144*, 014208.
2. Kinney, C.C.; Pytlewski, K.R.; Khachaturyan, A.G.; Morris, J.W., Jr. The microstructure of lath martensite in quenched 9Ni steel. *Acta Mater.* **2014**, *69*, 372–385. [CrossRef]
3. Sun, C.; Fu, P.X.; Liu, H.W.; Liu, H.H.; Du, N.Y. Effect of tempering temperature on the low temperature impact toughness of 42CrMo4-V steel. *Metals Open Access Metall. J.* **2018**, *8*, 232. [CrossRef]
4. Shibataa, A.; Nagoshi, T.; Sone, M.; Morito, S.; Higo, Y. Evaluation of the block boundary and sub-block boundary strengths of ferrous lath martensite using a micro-bending test. *Mater. Sci. Eng. A* **2010**, *527*, 7538–7544. [CrossRef]
5. Morito, S.; Tanaka, H.; Konishi, R.; Furuhara, T.; Maki, T. The morphology and crystallography of lath martensite in Fe-C alloys. *Acta Mater.* **2003**, *51*, 1789–1799. [CrossRef]
6. Morito, S.; Huang, X.; Furuhara, T.; Maki, T.; Hansen, N. The morphology and crystallography of lath martensite in alloy steels. *Acta Mater.* **2006**, *54*, 5323–5331. [CrossRef]
7. Grange, R.A. Strengthening steel by austenite grain refinement. *Trans. ASM* **1966**, *59*, 26–48.
8. Swarr, T.; Krauss, G. The effect of structure on the deformation of as-quenched and tempered martensite in a Fe-0.2 pct C alloy. *Metall. Trans. A* **1976**, *7A*, 41–48. [CrossRef]
9. Roberts, M.J. Effect of transformation substructure on the strength and toughness of Fe-Mn alloys. *Metall. Trans.* **1970**, *1*, 3287–3294.
10. Wang, C.F.; Wang, M.Q.; Shi, J.; Hui, W.J.; Dong, H. Effect of microstructure refinement on the strength and toughness of low alloy martensitic steel. *J. Mater. Sci. Technol.* **2007**, *23*, 659–664.
11. Morito, S.; Yoshida, H.; Maki, T.; Huang, X. Effect of block size on the strength of lath martensite in low carbon steels. *Mater. Sci. Eng. A* **2006**, *438–440*, 237–240. [CrossRef]
12. Zhang, C.Y.; Wang, Q.F.; Ren, J.X.; Li, R.X.; Wang, M.Z.; Zhang, F.C.; Yan, Z.S. Effect of microstructure on the strength of 25CrMo48V martensitic steel tempered at different temperature and time. *Mater. Des.* **2012**, *36*, 220–226. [CrossRef]
13. Long, S.L.; Liang, Y.L.; Jiang, Y.; Liang, Y.; Yang, M.; Yi, Y.L. Effect of quenching temperature on martensite multi-level microstructures and properties of strength and toughness in 20CrNi2Mo steel. *Mater. Sci. Eng. A* **2016**, *676*, 38–47. [CrossRef]
14. Li, S.C.; Zhu, G.M.; Kang, Y.L. Effect of substructure on mechanical properties and fracture behavior of lath martensite in 0.1C-1.1Si-1.7Mn steel. *J. Alloy. Compd.* **2016**, *675*, 104–115. [CrossRef]
15. Smith, D.W.; Hehemann, R.F. Influence of structural parameters on the yield strength of tempered martensite and lower bainite. *JISI* **1971**, 476–481.
16. Kim, B.; Boucard, E.; Sourmail, T.; Martín, D.S.; Gey, N.; Rivera-Díaz-del-Castillo, P.E.J. The influence of silicon in tempered martensite: Understanding the microstructure–properties relationship in 0.5–0.6 wt.% C steels. *Acta Mater.* **2014**, *68*, 169–178. [CrossRef]
17. Galindo-Nava, E.I.; Rivera-Díaz-del-Castillo, P.E.J. A model for the microstructure behaviour and strength evolution in lath martensite. *Acta Mater.* **2015**, *98*, 81–93. [CrossRef]
18. Wang, J.; Xu, Z.; Lu, X. Effect of the Quenching and Tempering Temperatures on the Microstructure and Mechanical Properties of H13 Steel. *J. Mater. Eng. Perform* **2020**. [CrossRef]
19. Speich, G.R.; Warlimont, H. Yield strength and transformation substructure of low-carbon martensite. *J. Iron Steel Res.* **1968**, *206*, 385–394.
20. Su, T.Y.H. Effect of lath martensite morphology on the mechanical properties steel. *Heat. Treat.* **2009**, *3*, 1–24.
21. Wang, S.X. *Metal Heat Treatment Principles and Process*; Harbin Industrial of Technology Press: Harbin, China, 2009.
22. Chen, J.; Lv, M.Y.; Tang, S.; Liu, Z.Y.; Wang, G.D. Influence of cooling paths on microstructural characteristics and precipitation behaviors in a low carbon V–Ti microalloyed steel. *Mater. Sci. Eng. A* **2014**, *594*, 389–393. [CrossRef]
23. Yen, H.W.; Chen, P.Y.; Huang, C.Y.; Yang, J.R. Interphase precipitation of nanometer-sized carbides in a titanium–molybdenum-bearing low-carbon steel. *Acta Mater.* **2011**, *59*, 6264–6274. [CrossRef]
24. Halfa, H. Recent trends in producing ultrafine grained steels. *J. Miner. Mater. Charact. Eng.* **2014**, *2*, 428–469. [CrossRef]
25. Cheng, X.Y.; Zhang, H.X.; Li, H.; Shen, H.P. Effect of tempering temperature on the microstructure and mechanical properties in mooring chain steel. *Mater. Sci. Eng. A* **2015**, *636*, 164–171. [CrossRef]

26. Peng, H.L.; Hu, L.; Ngai, T.W.; Li, L.J.; Zhang, X.L.; Xie, H.; Gong, W.P. Effects of austenitizing temperature on microstructure and mechanical property of a 4-GPa-grade PM high-speed steel. *Mater. Sci. Eng. A* **2018**, *719*, 21–26. [CrossRef]

27. Yong, Q.L. *Second Phases in Structural Steels*; Metallurgical Industry Press: Beijing, China, 2006.

28. HajyAkbary, F.; Sietsma, J.; B€ottger, A.J.; Santofimia, M.J. An improved X-ray diffraction analysis method to characterize dislocation density in lath martensitic structures. *Mater. Sci. Eng. A* **2015**, *639*, 208–218. [CrossRef]

29. Williamson, G.K.; Smallman, R.E. Dislocation densities in some annealed and cold-worked metals from measurements on the x-ray Debye-Scherrer spectrum. *Philos. Mag.* **1955**, *1*, 34–46. [CrossRef]

30. Wang, C.F. Study of Microstructure Control Unit on Strength and Toughness in Low Alloy Martensite Steel. Ph. D Thesis, Central Iron & Steel Research Institute, Beijing, China, 2008.

31. Shibata, A.; Nagoshi, T.; Sone, M.; Morito, S.; Higo, Y. Micromechanical characterization of deformation behavior in ferrous lath martensite. *J. Alloys. Compd.* **2013**, *577*, S555–S558. [CrossRef]

32. Lim, S.; Shin, C.; Heo, J.; Kim, S.; Jin, H.; Kwon, J.; Guim, H.; Jang, D. Micropillar compression study of the influence of size and internal boundary on the strength of HT9 tempered martensitic steel. *J. Nucl. Mater.* **2018**, *503*, 263–270. [CrossRef]

33. Mine, Y.; Hirashita, K.; Takashima, H.; Matsuda, M.; Takashima, K. Micro-tension behaviour of lath martensite structures of carbon steel. *Mater. Sci. Eng. A* **2013**, *560*, 535–544. [CrossRef]

34. Du, C.; Hoefnagels, J.P.M.; Vaes, R.; Geers, M.G.D. Block and sub-block boundary strengthening in lath martensite. *Scr. Mater.* **2016**, *116*, 117–121. [CrossRef]

35. Chen, S.J.; Yu, Q. The role of low angle grain boundary in deformation of titanium and its size effect. *Scr. Mater.* **2019**, *163*, 148–151. [CrossRef]

36. Hutchinson, B.; Hagström, J.; Karlsson, O.; Lindell, D.; Tornberg, M.; Lindberg, F.; Thuvander, M. Microstructures and hardness of as-quenched martensites (0.1–0.5% C). *Acta Mater.* **2011**, *59*, 5845–5858. [CrossRef]

37. Takahashi, J.; Ishikawa, K.; Kawakami, K.; Fujioka, M.; Kubota, N. Atomic-scale study on segregation behavior at austenite grain boundaries in boron- and molybdenum-added steels. *Acta Mater.* **2017**, *133*, 41–54. [CrossRef]

38. Miyamoto, G.; Goto, A.; Takayama, N.; Furuhara, T. Three-dimensional atom probe analysis of boron segregation at austenite grain boundary in a low carbon steel - Effects of boundary misorientation and quenching temperature. *Scr. Mater.* **2018**, *154*, 168–171. [CrossRef]

39. Waseda, O.; Veiga, R.G.; Morthomas, J.; Chantrenne, P.; Becquart, C.S.; Ribeiro, F.; Jelea, A.; Goldenstein, H.; Perez, M. Formation of carbon Cottrell atmospheres and their effect on the stress field around an edge dislocation. *Scr. Mater.* **2017**, *129*, 16–19. [CrossRef]

Article

Microstructural Stability and Softening Resistance of a Novel Hot-Work Die Steel

Ningyu Du [1,2], Hongwei Liu [1,*], Paixian Fu [1,*], Hanghang Liu [1], Chen Sun [1,2], Yanfei Cao [1,*] and Dianzhong Li [1]

[1]   Shenyang National Laboratory for Materials Science, Institute of Metal Research, Chinese Academy of Sciences, Shenyang 110016, China; nydu16s@imr.ac.cn (N.D.); hhliu@imr.ac.cn (H.L.); csun15s@imr.ac.cn (C.S.); dzli@imr.ac.cn (D.L.)
[2]   School of Materials Science and Engineering, University of Science and Technology of China, Hefei 230026, China
*   Correspondence: hwliu@imr.ac.cn (H.L.); pxfu@imr.ac.cn (P.F.); yfcao10s@imr.ac.cn (Y.C.)

Received: 7 March 2020; Accepted: 24 March 2020; Published: 25 March 2020

Abstract: A novel hot-work die steel, named 5Cr5Mo2, was designed to obtain superior thermal stability. The proposed alloy is evaluated in terms of its hardness, microstructure, and tempering kinetics. Compared with the commonly used H13 steel, the softening resistance of the designed steel is superior. Based on SEM and transmission electron microscopy (TEM) observations, a higher abundance of fine molybdenum carbides precipitate in 5Cr5Mo2 steel. Strikingly, the coarseness rate of the carbides is also relatively low during the tempering treatment. Moreover, owing to their pinning effect on dislocation slip, the dislocation density of the 5Cr5Mo2 steel decreases more slowly than that of the H13 steel. Furthermore, a mathematical softening model was successfully deduced and verified by analyzing the tempering kinetics. This model can be used to predict the hardness evolution of the die steels during the service period at high temperature.

Keywords: hot-work die steel; thermal stability; carbide; dislocation; tempering kinetics

1. Introduction

Hot-work die steels are steel alloys that are suitable for manufacturing dies for the hot deformation of metals, such as hot extrusion dies, hot-forging dies, hot upsetting dies, and die-casting dies. Hot-work die steel alloys are commonly used to produce tools that are exposed to high temperatures and are usually alloyed with strong carbide forming elements such as Cr, Mo, and V [1,2]. Hot-forming is a reliable and relatively economical manufacturing process for metal parts, which relies on high-quality hot-work dies. During hot-forming, regions of the tools may attain high temperatures or pressures, especially regions of the die tool that exhibit a small radii in the die cavities [3]. Because hot-forming dies are exposed to high temperatures and pressures for prolonged periods, the die materials are required to exhibit high strength, hardness, particularly high thermal strength, thermal fatigue, toughness, and wear resistance. The risk of encountering various failure modes drastically increases the probability of the properties of the selected tool material to not meet the service conditions. Consequently, the manufacturing costs can be significantly increased [4]. A general requirement of steel alloys with better thermal stability is that their microstructure and hardness are not likely to rapidly change at high temperatures. Thus, the thermal stability of an alloy, which is a crucial property influencing the service life of hot-work die steel alloys, must be considered before the alloy is used for engineering applications.

Earlier investigations [5–7] of hot-work die steel alloys have shown that the thermal stability during tempering is strongly associated with the microstructure and the alloy carbide precipitates. The typical microstructure of hot-work die steel alloys after conventional quenching and tempering procedures consists of tempered martensite with solid solution of carbon, high dislocation density, and fine alloy carbide precipitates. Nevertheless, a prolonged tempering treatment results in the softening of steel alloys, which is mainly caused by the coarsening of the fine alloy carbides, reduction of the dislocation density, and recovery of the martensitic structure [6,8,9]. By observing the microstructure of low-carbon martensitic steels, A. Arlazarov [10] showed the precipitation of carbides caused by tempering. The amount and size of precipitates is significantly influenced by carbide-forming elements (W, Mo, Nb, V) [11]. D. Delagnes et al. [12] explained the influence of the silicon content on the amount of carbides in a 5%Cr tempered martensitic steel.

Hot-forming dies are used in high-temperature environments, which can be regarded, to some extent, as a prolonged tempering process. During tempering, carbon atoms eject from the supersaturated martensite, consequently causing the reduction of the strength of steel alloys [13]. The martensite tempering process is influenced by tempering temperature [14], and tempering time [15,16]. The effects of chemical composition and the heat treatment parameters on the microstructure and properties of die steel alloys have been extensively studied [17–19]. However, the microstructural stability and softening resistance of hot-work die steel alloys during long-term operation in the high-temperature situation have seldom been focused on.

In this study, we designed a novel 5Cr5Mo2 steel alloy based on the most common hot-work H13 die steel. The thermal stability of the proposed alloy was systemically determined based on its microstructure, hardness, and tempering kinetics. The main factors affecting the softening and the mechanism of the softening during tempering were clarified. Most importantly, a mathematical model describing the softening behavior of the alloy was successfully built.

2. Materials and Experiments

2.1. Materials and Heat Treatment

5Cr5Mo2 and H13 hot-work die steel samples were produced using vacuum induction melting. The chemical composition of the two steel alloys is given in Table 1. All samples for this experiment were obtained by wire electrical machining from a single ingot after forging along the three orthogonal directions.

Table 1. Measured chemical composition of two alloys, in wt%, balance Fe.

Steel	C	Cr	Mo	V	Si	Mn
5Cr5Mo2	0.50	5.14	2.48	0.51	0.20	0.50
H13	0.32	5.05	1.35	0.90	0.97	0.32

In order to obtain the service microstructure and set the hardness of two steel alloys to a similar value, the samples used for thermal stability tests were subjected to a pre-heat-treatment process consisting of quenching and tempering. Our previous experiments showed that, the hardness and Charpy un-notched impact energy of 5Cr5Mo2 were the highest with an austenitizing temperature of 1050 °C, and its Charpy V-notch impact energy is slightly smaller than that at the lower temperatures, as shown in Figure 1a–c. When a tempering temperature of 600 °C is then applied, the hardness reaches 49.8 HRC (see Figure 1d), which is close to the follow-up experimental requirements, making the hardness of two steel alloys to a similar value and obtaining an excellent performance in the 5Cr5Mo2 steel. Hence, in this study, the 5Cr5Mo2 steel was heated at an austenitizing temperature of 1050 °C for 1 h and oil quenched. Subsequently, the samples were tempered at 600 °C for 2 h, followed by air cooling (Figure 2a). In terms of the H13 steel, the common heat treatment parameters were used as shown in Figure 2b. For instance, the H13 steel was heated at an austenitizing temperature of 1030 °C for one hour and oil quenched. Subsequently, the samples were tempered at 580 °C for 2 h, and then air-cooled. The tempering process was repeated two times. It should be stressed that, the choice of a lower quenching temperature (1030 °C) for H13 steel is because that, more undissolved VC carbides play an important role in refining grains during quenching. Such heat treatment process is commonly used in engineering practice.

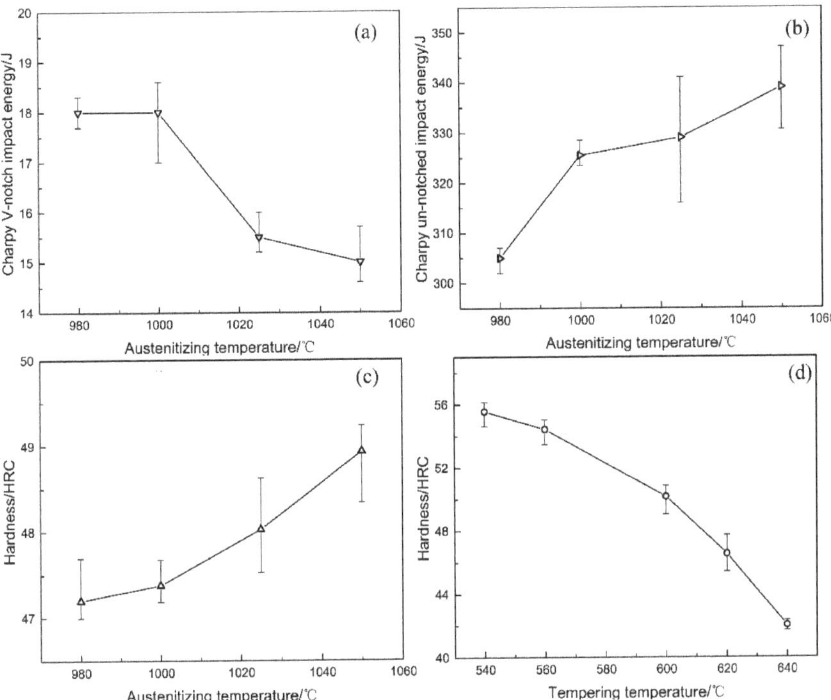

Figure 1. The effect of different austenitizing temperatures on (**a**) Charpy V–notch impact energy; (**b**) Charpy un-notched impact energy and (**c**) hardness; and (**d**) hardness value as a function of the tempering temperatures of the 5Cr5Mo2 steel with an austenitizing temperatures of 1050 °C.

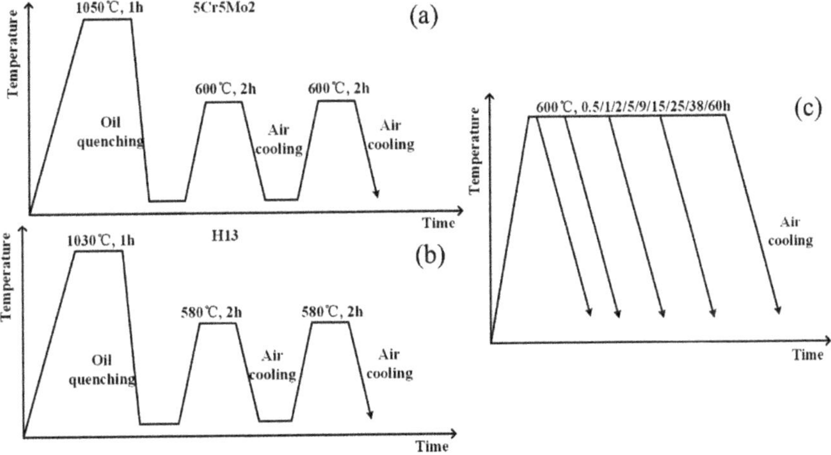

Figure 2. The pre-heat-treatment process of (**a**) 5Cr5Mo2 steel and (**b**) H13 steel; (**c**) The additional tempering process.

To obtain the evolution of the thermal stability in high-temperature service conditions—such as hardness and characteristics of the microstructure—an additional tempering was performed at 600 °C which is the typical working temperature of forging dies. The additional tempering process is shown in Figure 2c. This secondary tempering lasted various durations: {0.5, 1, 2, 5, 9, 15, 25, 38, 60} h. The tempering was followed by air cooling. Samples with dimensions of 12 mm × 12 mm × 10 mm were prepared for thermal stability tests. In order to prevent oxidation during the heat treatment, all samples were sealed in quartz tubes filled with argon.

2.2. Mechanical Tests

Rockwell hardness was measured in a Wilson Rockwell device (LCR–500, LECO Company) according to the standard ASTM E18-2017. Samples with dimensions of 12 mm × 12 mm × 10 mm prior to the pre-treatment were prepared through tempering, whose duration varied among the samples to investigate the influence of the tempering duration.

2.3. Microstructure Observations

The microstructure characteristics of the 5Cr5Mo2 and H13 specimens were investigated by scanning electron microscopy (Inspect F50, FEI) and transmission electron microscopy (TEM) (Tecnai Spirit TEM T12, FEI and Tecnai G2 20, FEI). The transmission electron microscopy samples with a diameter of 3 mm and thickness of 50 μm were thinned by twin-jet electropolishing in a 6 vol.% perchloric acid/ethanol solution at a voltage of 25 V and a temperature between −30 °C and −25 °C.

2.4. X-ray Diffraction Analysis

X-ray diffraction (XRD) (D8 Advance, Bruker) analysis was performed using incidence angles of 40°–100° for the solid bulk samples and incidence angles of 20°–85° for the carbide powder extracted from the solid samples at room temperature using Cu-Kα radiation. The solid bulk samples used for the dislocation density measurements were prepared by the electro-polishing method in order to prevent the presence of residual stresses that appeared during mechanical polishing. The second phase particles of the two steel alloys were characterized by electrolytic extraction in a 10 vol.% hydrochloric acid/methanol solution. The samples for extraction were taken after the pre-heat-treatment. The dislocation density and types of second phase particles were determined by analyzing the XRD patterns via the MDI Jade 6.5 software. The microstrain and the full width at half-maximum intensity (FWHM) of the selected peaks were determined by the Williamson–Hall method [20]. The dislocation density values were calculated from the microstrain by the Williamson–Hall equation [5,21–23].

3. Results and Discussion

3.1. Alloy Design

V and Mo are strong carbide forming elements which are commonly used for tool steel alloys. Both V and Mo are effective strengthening elements. Molybdenum plays an important role in retarding the coarsening of nano-sized carbides and inhibiting the annihilation of dislocations at elevated tempering temperatures [24,25]. Steel alloys with Mo additions exhibit longer retention of their mechanical properties during high-temperature aging [26–28]. Hence, during the design of the proposed steel alloy, the Mo content of the steel was adjusted from 1.35 wt.% to 2.48 wt.% to obtain finer carbides and enhanced characteristics. On the contrary, V-rich MC particles can coarsen during hot rolling or during the subsequent annealing heat treatment, thus, leading to the significant degradation of the strength of the materials. Moreover, it was shown that large primary vanadium carbide particles are formed in H13 steel, which is detrimental to the steel's properties. Considering the detrimental effect and material cost, the V content of the steel was reduced.

Carbon atoms present in steels either form a solid solution in the steel matrix or participate in the formation of carbides. Carbon increases the strength and hardness of the steel but reduces its toughness. Therefore, the carbon content of the proposed alloy was increased to ensure sufficient hardness in high-temperature environments, while keeping it appropriately low to avoid a serious decrease in toughness. Silicon also significantly affects the properties of steel by increasing the quantity and stability of the retained austenite [24]. However, the transformation of austenite to martensite would deteriorate the dimensional accuracy of the produced dies and the machining performance of the steels [29]. Thus, a high retained austenite fraction is generally avoided in hot-work die steel alloys. Consequently, the Si content of the proposed 5Cr5Mo2 steel is moderately reduced. The chemical composition of the two target alloys considered is shown in Table 1.

The Thermo-Calc software was used to carry out thermodynamic calculations to guide the alloy design and to plan the follow-up experimental study. The subsequent thermal stability tests were carried out at the temperature of 600 °C, and the liquidus temperature of the two considered steel alloys is below 1500 °C. Consequently, the temperature range selected for the calculations was 400–1500 °C. The results of the calculations are shown in Figure 3. The main equilibrium precipitation phases in the 5Cr5Mo2 steel are MC, M_2C, M_7C_3, M_6C, and $M_{23}C_6$. There are the same types of equilibrium precipitation phases that are present in the H13 steel. The temperature-dependent behavior of certain precipitation phases exhibits sudden changes. This is caused by the change of carbide types, which are present in the steel in a non-equilibrium state for a prolonged duration. After keeping the H13 steel at the tempering temperature of 600 °C, MC, $M_{23}C_6$, and M_6C exist stably, while M_2C and M_7C_3 dissolve or transform into other phases. On the contrary, the stable phases in the 5Cr5Mo2 steel are MC, M_2C, and $M_{23}C_6$, while M_7C_3 and M_6C are unstable at 600 °C. The composition of these phases was also calculated by the Thermo-Calc software. The variation curves of the constituting elements (in at. %) of the relevant phases are shown in Figure 4 for a temperature range of 400–1500 °C: MC, M_2C, $M_{23}C_6$, and M_6C are V-rich, Mo-rich, Cr-rich, and Mo-rich phases, respectively. There are four stable phases in the 5Cr5Mo2 steel at 600 °C: α, MC, $M_{23}C_6$, and M_2C with a mole fraction of 90.0%, 0.4%, 8.7%, and 0.9%, respectively. Conversely, the mole fractions of α, MC, $M_{23}C_6$, and M_6C in the H13 steel are 94.2%, 4.5%, 1.1%, and 0.2%, respectively. These results are consistent with the intended alloy design.

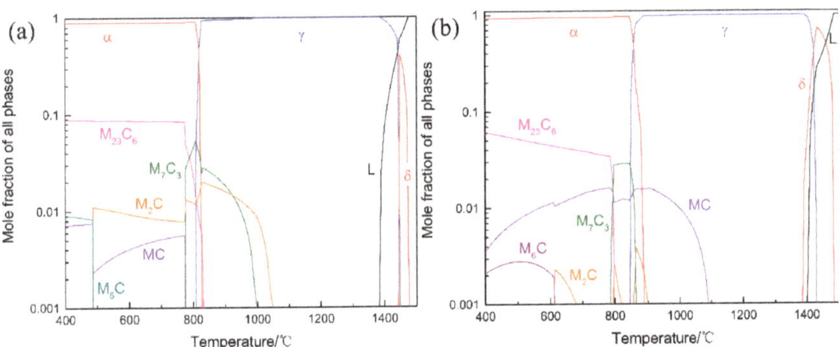

Figure 3. Results of the thermodynamic calculation of the equilibrium precipitation phases of the (**a**) 5Cr5Mo2 steel; and (**b**) H13 steel.

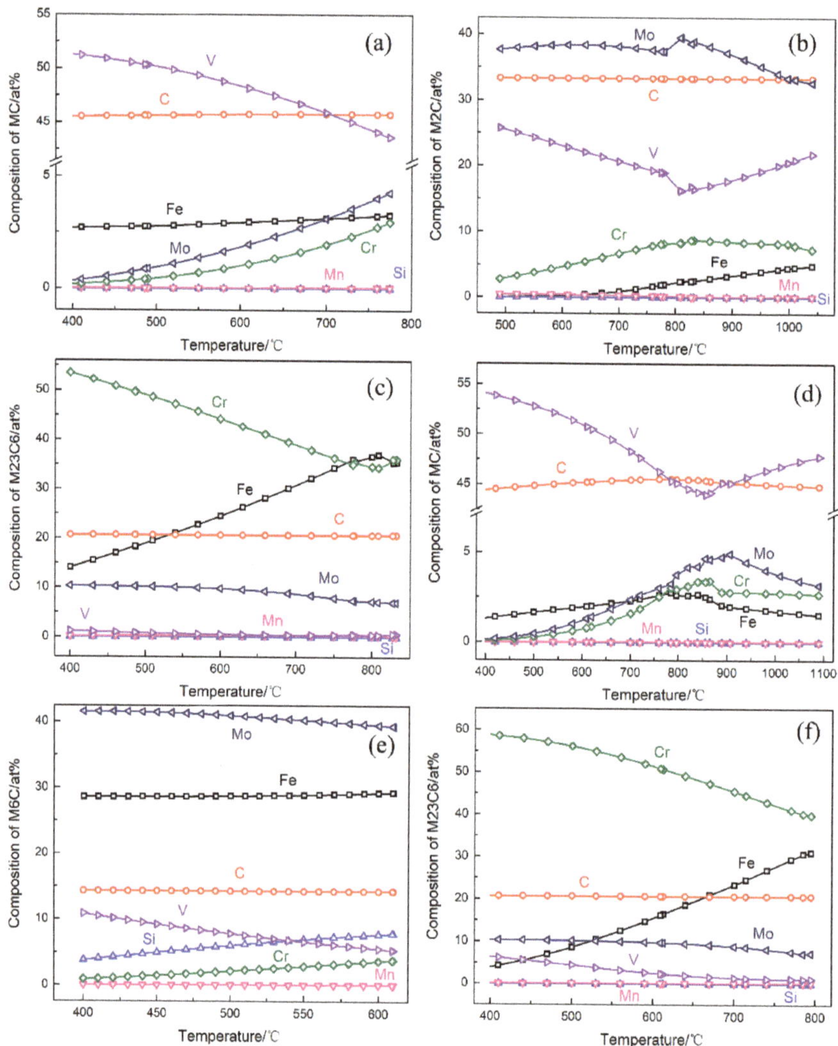

Figure 4. Equilibrium composition of the (**a**) MC, (**b**) M_2C, and (**c**) $M_{23}C_6$ phases in the 5Cr5Mo2 steel and (**d**) MC, (**e**) M_6C, (**f**) $M_{23}C_6$ phase in the H13 steel.

3.2. Mechanical Properties

The hardness of the 5Cr5Mo2 and H13 steel alloys as a function of the tempering duration is shown in Figure 5. The hardness of both steel alloys decreases as the tempering duration increases. Nevertheless, the hardness of the H13 steel decreases faster than that of the 5Cr5Mo2 steel. The hardness of the 5Cr5Mo2 steel decreases from 49.8 HRC to 36.4 HRC. At the same time, the hardness of the H13 steel decreases from 49.3 HRC to 27.3 HRC after 60 h of tempering at 600 °C. In addition, a sharp decrease in hardness of the two steel alloys occurs during the initial stage of tempering. On the contrary, the softening rate decreases with the increased tempering duration. For comparison, the softening rate of the H13 steel is faster than that of the 5Cr5Mo2 steel. However, the softening rate of both alloys decreases with the increasing tempering duration. That is, the softening resistance of the 5Cr5Mo2 steel is higher than that of the H13 steel and the designed steel exhibits better thermal stability.

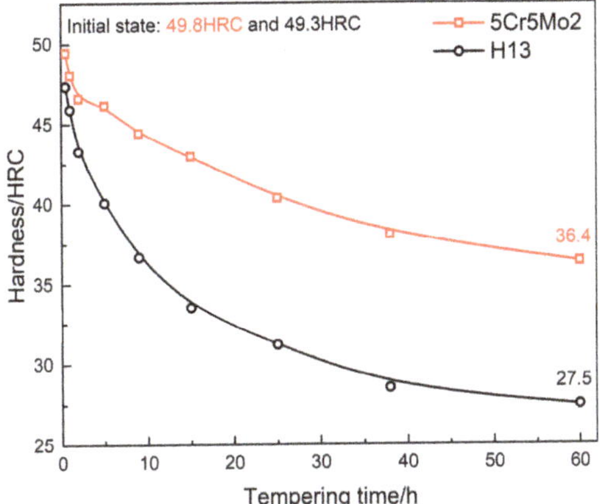

Figure 5. Hardness of the 5Cr5Mo2 and H13 steel alloys as functions of the tempering duration.

3.3. Microstructure

The softening of the 5Cr5Mo2 and H13 steel alloys during tempering is strongly linked to their microstructure. Most alloying elements dissolve into the matrix during quenching and separate from the martensite matrix during tempering, which plays a significant role in dispersion strengthening. Correspondingly, the microstructure of the two steel alloys after the pre-heat-treatment without additional tempering mainly consists of tempered martensite and a small number of fine alloy carbides, as shown in Figure 6a,d. After tempering for 25 h and 60 h, the fine carbides appear significantly coarsened, and the martensite matrix is recovered in both steel alloys, as shown in Figure 6b,c,e,f. However, by comparing the structures of the two steel alloys, it can be observed that the modification of the tempered martensitic lath of the 5Cr5Mo2 alloy is relatively small compared with that of the H13 steel. Thus, it can be concluded that the martensite of the designed steel is more stable than that of the H13 steel.

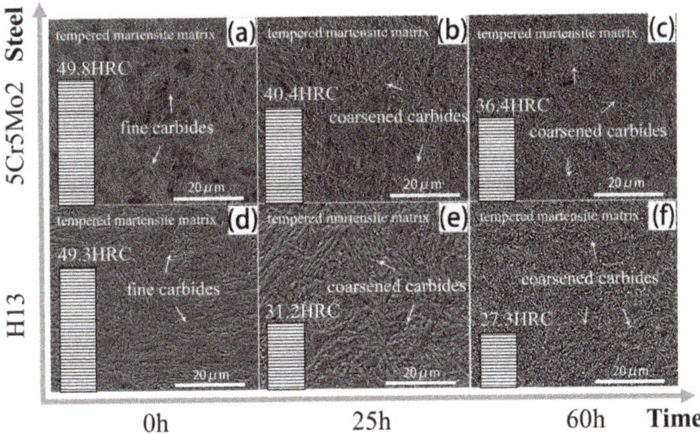

Figure 6. Microstructure and hardness after a tempering lasting (a) 0 h, (b) 25 h, and (c) 60 h for the 5Cr5Mo2 steel; and (d) 0 h, (e) 25 h, (f) 60 h for the H13 steel.

In essence, steel softening is the degradation of multiple reinforcing effects, such as solid solution strengthening, and precipitation strengthening after the tempering process. The strengthening mechanisms of martensite achieved by quenching have been already summarized elsewhere [30], including the substitutional and interstitial solid solutions, dislocation strengthening, segregation of carbon atoms, and precipitation of carbides. After the tempering treatment, the strengthening effect of the martensite structure is significantly reduced [24,30]. The softening of hot-work die steel alloys mainly depends on the recovery of dislocations and the coarsening of alloy carbides at high temperatures [30]. Kelly et al. [31] considered that although the segregation of carbon reduces the stress field of the dislocations, it plays a more significant role in pinning dislocations, consequently increasing the strength of the alloy after tempering. On the contrary, carbides precipitate at dislocations, which is why the strength of martensite is shown to increase at room temperature or just above room temperature [30]. Thus, the factors influencing martensite stability and the difference between these factors in the two steel alloys were analyzed based on the characteristics of carbides and dislocations.

High-magnification SEM and TEM observations were performed to obtain more detailed information about the steels' microstructures. Figure 7 shows the microstructure of both steel alloys after no additional tempering. The matrices of both steel alloys are tempered martensite. In the 5Cr5Mo2 steel, the most abundant carbides are elliptical-shaped vanadium carbides and fine needle-shaped molybdenum carbides, as shown in Figure 7b,c. Conversely, there are some large size primary carbides found in the H13 steel, as shown in Figure 7f,g. Those carbides are vanadium carbides identified by their diffraction (Figure 7h) and energy spectra (Figure 7i). The second carbides are mainly vanadium carbides. XRD measurements were also performed on carbide powders extracted from the solid samples. The XRD results (Figure 8) were consistent with the TEM observations.

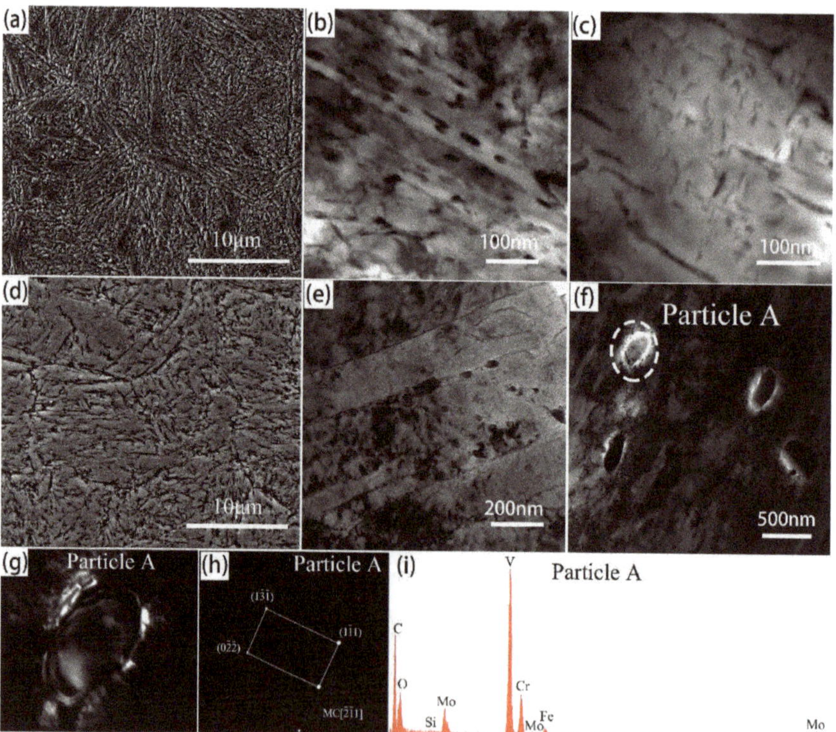

Figure 7. Microstructure of both steel alloys prior to additional tempering: (**a**) microstructure of the 5Cr5Mo2 steel observed by SEM; (**b**) elliptical-shaped secondary carbides and (**c**) fine needle-shaped secondary carbides of the 5Cr5Mo2 steel observed by TEM; (**d**) microstructure of the H13 steel observed by SEM; (**e**) secondary carbides and (**f**) large-size primary carbides of the H13 steel observed by TEM; (**g**) morphology, (**h**) diffraction, and (**i**) energy spectra of particle A.

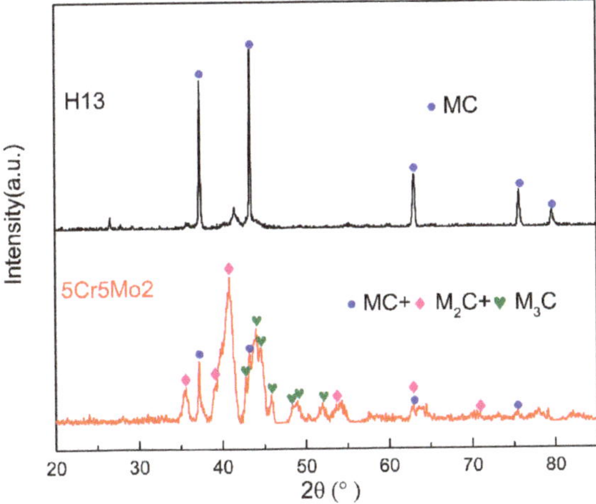

Figure 8. X-ray diffraction (XRD) patterns of the carbide powder extracted from the 5Cr5Mo2 and H13 steels.

The fine molybdenum and vanadium carbides are conducive to the thermal stability, but the large size primary carbides are detrimental for the toughness. The designed steel, with higher molybdenum and lower vanadium content, effectively avoids the formation of large size primary carbides and promotes the formation of fine molybdenum carbides. Moreover, molybdenum carbides are usually fine and coarsen slowly in steel alloys [32]. Based on the comparison of the carbides present in the two steel alloys (Figure 7), it can also be concluded that more carbides precipitate in the 5Cr5Mo2 steel because of its higher carbon content.

The evolutions of the tempered martensite and carbides after additional tempering duration of 60 h are shown in Figure 9 for both steel alloys. Figure 9a,d show the widened martensite and coarsened carbides in both steel alloys observed by SEM. Although martensite recovery and carbide coarsening occur in both alloys, a difference between the designed steel and the H13 steel can be observed. The carbides in the 5Cr5Mo2 steel are predominantly elliptical-shaped and bar-like carbides, as observed by TEM (Figure 9b,c). An elliptical-shaped carbide (particle B) was further investigated by its diffraction (Figure 9g) and energy spectra (Figure 9h). The measurements suggest that particle B is a Cr-rich M$_{23}$C$_6$ type carbide. TEM observations show that large-size carbides are abundant in the H13 steel, as shown in Figure 9e,f. The size of the largest observed secondary carbide exceeded 200 nm. The carbides are mainly vanadium carbides, as identified by their diffraction (Figure 9i) and energy spectra (Figure 9j).

By comparing the martensite and carbides with additional tempering for 0 h (Figure 7) and 60 h (Figure 9), the martensite clearly recovers while the carbides coarsen to a higher degree after the longer tempering. Carbide formation requires the diffusion of alloy elements which allows the precipitation of carbides from the steel during tempering. With the increase of the tempering time, subsequent changes in the morphology of the carbides occur according to the Ostwald ripening process, where smaller carbides dissolve into the matrix, providing carbon for the selective growth of larger carbides [19].

Figure 10 shows a schematic illustration of the evolution of the martensite hierarchical structures. During tempering, carbides precipitate and grow, while martensitic lath and dislocations recover. The main factors affecting the decrease in hardness of both steel alloys are martensite, carbides, and dislocations. Thus, the dislocation density of both steel alloys was calculated based on experimental data measured by X-ray diffraction, which is presented in the following section.

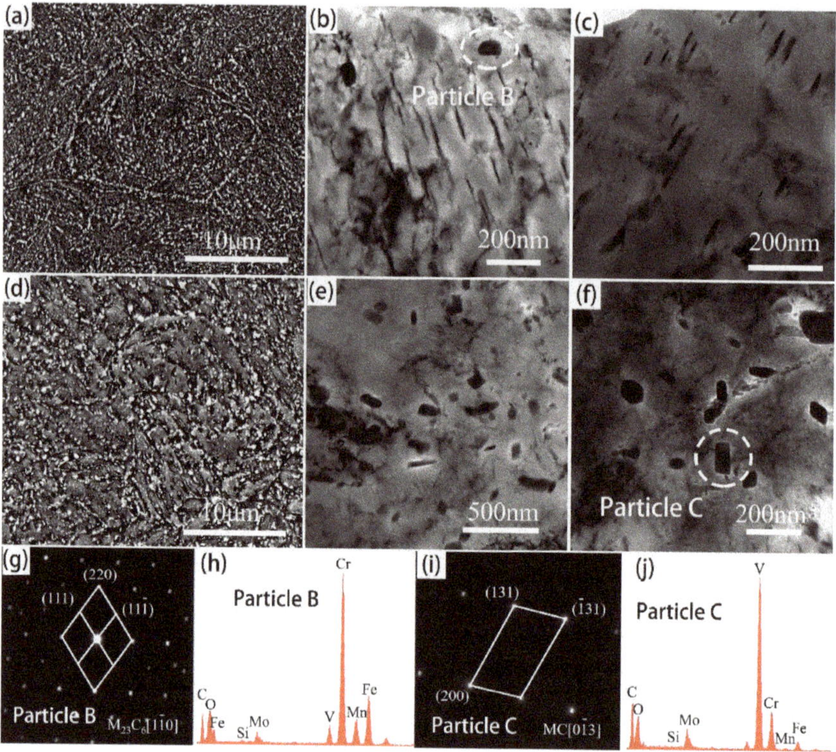

Figure 9. Microstructure of both steel alloys after tempering for 60 h: (**a**) microstructure of the 5Cr5Mo2 steel observed by SEM; (**b**) elliptical-shaped secondary carbides and (**c**) bar-like secondary carbides in the 5Cr5Mo2 steel observed by TEM; (**d**) microstructure of the H13 steel observed by SEM; (**e**) large-size and (**f**) fine secondary carbides in the H13 steel observed by TEM; (**g**) diffraction and (**h**) energy spectra of particle B; (**i**) diffraction and (**j**) energy spectra of particle C.

3.4. Dislocation Density

Martensite is a supersaturated solid solution of carbon in α-Fe, where the interstitial carbon atoms cause an asymmetric distortion of the lattice. The non-uniform stress field of the asymmetric distorted centers (carbon atoms) strongly interacts with the dislocations. Thus, the non-uniform stress field strongly hinders the movement of dislocations, resulting in the broadening of X-ray diffraction lines. Therefore, changes in the X-ray diffraction peaks can be used to characterize the dislocation densities of the alloys. The dislocation density ρ was calculated from the average microstrain values e according to [33]:

$$\rho = 14.4\frac{e^2}{b^2},\tag{1}$$

where b is the magnitude of the Burger's vector ($b = \sqrt{3}a/2$ for the bcc structure where a is the lattice parameter). The microstrain e can be calculated using the following Williamson–Hall equation [34]:

$$\frac{\delta_{hkl}\cos\theta_{hkl}}{\lambda} = \frac{1}{D} + \frac{4e\sin\theta_{hkl}}{\lambda},\tag{2}$$

where θ_{hkl} is the diffraction angle, D is the mean size of the coherently diffracting domains, and λ is X-ray wavelength. In addition, δ_{hkl} was measured on a scale of $2\theta_{hkl}$ and analyzed among four

XRD peaks. The value of the microstrain e was obtained from a slop of a straight line, which could be generated by plotting $\delta_{hkl}\cos\theta_{hkl}/\lambda$ vs. $4\sin\theta_{hkl}/\lambda$.

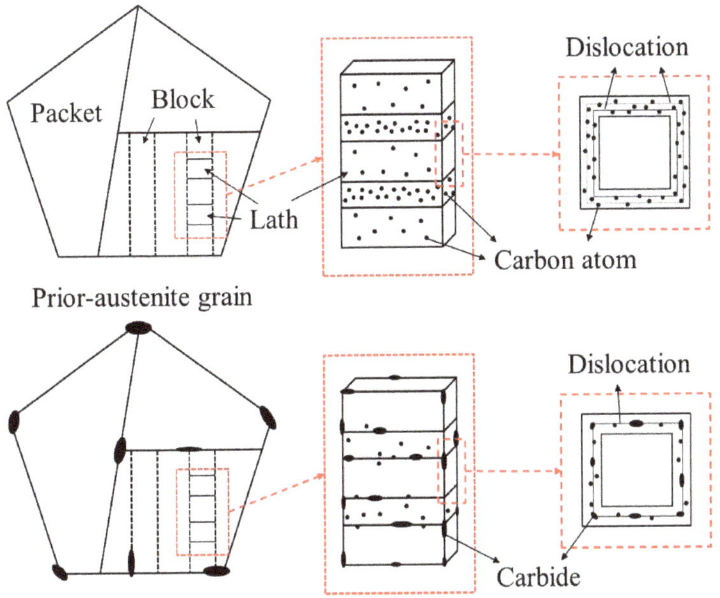

Figure 10. Schematic illustration of the evolution of martensite, carbides, and dislocations after a long-lasting tempering.

The dislocation density ρ can be calculated by Equation (1). The dislocation densities are plotted in Figure 11 as a function of the tempering duration. The dislocation density of the 5Cr5Mo2 specimens in their initial state (additional tempering for 0 h) is lower than that of the H13 steel. This is due to the different pre-heat-treatment processes used for the two steel alloys. The pre-heat-treatment temperatures of the 5Cr5Mo2 and H13 steel alloys were 600 °C and 580 °C, respectively. A higher tempering temperature results in a lower dislocation density. During the first 10 h of tempering, the dislocations recover, and the dislocation density sharply declines in both steel alloys. The subsequent decline is relatively slow. A schematic illustration of the dislocation changes after the tempering is also shown in Figure 10. The dislocation density of the 5Cr5Mo2 steel exceeds that of the H13 steel when both alloys are exposed to tempering exceeding 25 h. In general, the dislocation recovery rate of the 5Cr5Mo2 steel is lower than that of the H13 steel. This is because more fine precipitates along the lath boundaries and the lath result in more effective pinning dislocation slips.

3.5. Tempering Kinetics and the Softening Model

In the following section, a kinetic model is used to describe the solid transformation and recrystallization in the metallic materials. It is assumed that the solid-phase transformation is controlled by diffusion as proposed by Johnson and Mehl [35], and Avrami [36–38]. The model takes in the form of the Johnson–Mehl–Avrami equation as follows [7,39]:

$$\tau = 1 - \exp(-Dt^n), \tag{3}$$

where τ, t, and n are the tempering ratio, the heat preservation time of the temper, and the Avrami exponent, respectively. D could be determined according to the Arrhenius equation:

$$D = D_0 \exp\left(-\frac{Q}{RT}\right),\tag{4}$$

where D_0, Q, T and R are the pre-exponential constant, the activation energy of the tempering transformation, the tempering isothermal temperature in K, and the ideal gas constant $(8.314\ \mathrm{J\cdot mol^{-1}\cdot K^{-1}})$, respectively.

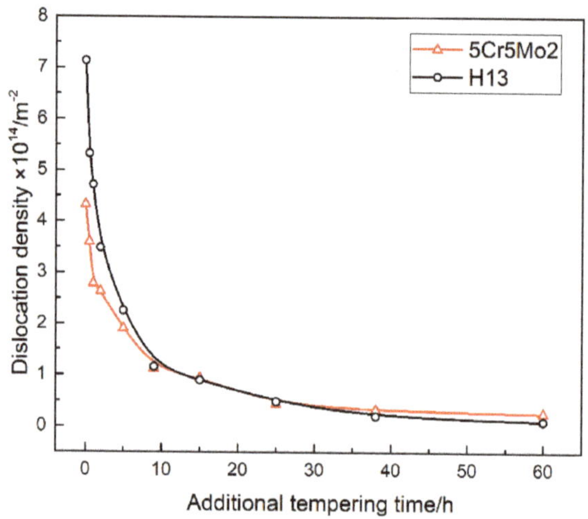

Figure 11. Dislocation density as a function of the tempering duration.

The following equation is used to describe the relationship between the tempering rate and the hardness [6]:

$$\tau = \frac{H_v - H_0}{H_\infty - H_0},\tag{5}$$

where H_0, H_∞, and H_v are the hardness in the quenched state, the hardness in the annealed state, and the hardness after tempering a duration of v, respectively.

By inserting Equation (5) into Equation (3), the relationship between the tempering hardness H_v, D, and the tempering time t can be written as

$$\ln\ln\frac{H_0 - H_\infty}{H_v - H_\infty} = n\ln t + \ln D.\tag{6}$$

Using the hardness measurements of the two steel alloys tempered for various durations, the values of the Avrami exponent n and $\ln D$ can be obtained according to Equation (6). The data of $\ln\ln\frac{H_0 - H_\infty}{H_v - H_\infty}$ and $\ln t$ were plotted and fitted by linear curves (Figure 12a). It should be noted that, a single linear curve cannot fit the data of the 5Cr5Mo2 steel well. Hence, two linear curves with different slopes were used. The Avrami exponent n of the H13 steel tempered at 600 °C was 0.36. Conversely, the n values of the 5Cr5Mo2 steel during the first stage of the tempering (t < 11.5 h) were 0.14, and subsequently increased to 0.30 during the second stage (t > 11.5 h). The value of n describes the growth mechanism of the coarsening of large spheroids [7]. Therefore, the growth mechanism of the coarsening of large spheroids of the two steel alloys differs. Hence, it is speculated that this is caused by the presence of different types of carbides in the two steel alloys.

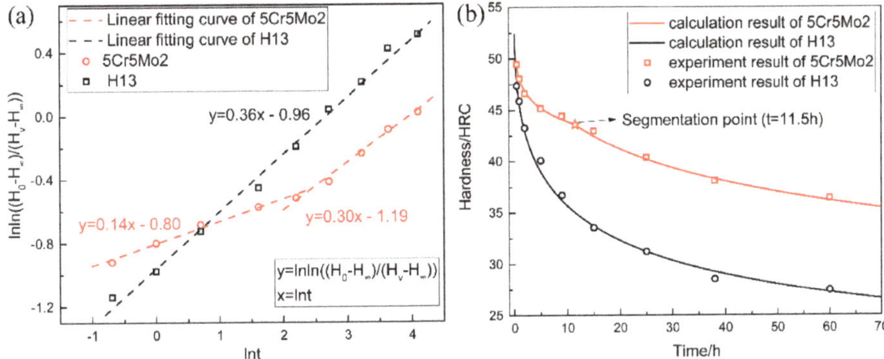

Figure 12. (**a**) Determination of the Avrami exponent n and $\ln D$ using curve fitting. (**b**) Softening curves and experimental hardness results of the 5Cr5Mo2 and H13 steel alloys.

The values of $\ln D$, which were obtained by extrapolating the curves to intersect the axis of ordinates, were -0.80 and -1.19 for the 5Cr5Mo2 steel and -0.96 for the H13 steel respectively. By transforming Equation (6), the tempering hardness can be expressed as

$$H_v = H_\infty + (H_0 - H_\infty)\exp(-Dt^n). \tag{7}$$

Inserting n, $\ln D$, H_∞, and H_0 into Equation (7), the softening equations of 5Cr5Mo2 and H13 can be obtained:

$$H_v = \begin{cases} 21.3 + 42\exp\!\left(-e^{-0.80}t^{0.14}\right), t < 11.5\ \text{h} \\ 21.3 + 42\exp\!\left(-e^{-1.19}t^{0.30}\right), t > 11.5\ \text{h} \end{cases} \tag{8}$$

$$H_v = 20.3 + 37\exp\!\left(-e^{-0.96}t^{0.36}\right). \tag{9}$$

The softening curves of the two steel alloys given by Equation (8) and Equation (9) are shown in Figure 12b. In addition, the measured hardness values are plotted in Figure 12b to verify the accuracy of the proposed equations. The calculated softening curves are generally consistent with the experimental results. Therefore, the softening equations can be used to predict the hardness of dies after prolonged usage at 600 °C.

4. Conclusions

In this study, a novel hot-work die steel of 5Cr5Mo2 with excellent thermal stability was designed. The mechanisms of the alloy's superior thermal stability were elucidated by analyzing its hardness, microstructure, and tempering kinetics. The main observation and conclusion can be summarized as follows:

(1) The designed steel exhibits a superior softening resistance as compared with the common H13 steel. After tempering for 60 h, the hardness of the 5Cr5Mo2 and H13 steels decreased from 49.8 HRC and 49.3 HRC to 36.4 HRC and 27.3 HRC, respectively.

(2) The initial microstructure of the designed steel without the additional tempering primarily consisted of tempered martensite and fine alloy carbides. The typical secondary carbides in the 5Cr5Mo2 steel were elliptical vanadium carbides and fine acicular molybdenum carbides.

(3) By the additional tempering, the smaller carbides were dissolved, and the coarsening of the selective carbide occurred. Martensites as well as dislocations were recovered in both steels. However, the initial characteristics of the martensite morphology were more pronounced in the 5Cr5Mo2 steel. Moreover, relatively finer carbides were retained in the 5Cr5Mo2 steel as compared to those in the H13 steel.

(4) Initially, the dislocation density of the 5Cr5Mo2 steel was lower than that of the H13 steel due to its different pre-heat-treatment process. Nevertheless, after being exposed to an additional tempering lasting over 25 h, owing to the dislocation pinning by precipitation, the dislocation density exceeded that of the H13 steel.

(5) By calculating the tempering kinetics of both steels, their softening equations were obtained and validated. The equations can be used to effectively predict the hardness of the dies after the prolonged service period at 600 °C.

Author Contributions: Conceptualization, N.D., Y.C. and H.L. (Hongwei Liu); methodology, N.D.; validation, N.D., Y.C. and H.L. (Hongwei Liu); formal analysis, N.D.; investigation, N.D., Y.C. and H.L. (Hongwei Liu); resources, H.L. (Hongwei Liu), P.F. and D.L.; data curation, N.D. and H.L. (Hongwei Liu); writing—original draft preparation, N.D.; writing—review and editing, N.D., Y.C., H.L. (Hongwei Liu), P.F., H.L. (Hanghang Liu) C.S.; visualization, N.D., Y.C. and H.L. (Hongwei Liu); supervision, D.L.; project administration, D.L.; funding acquisition, Y.C., H.L. (Hongwei Liu), P.F., H.L. (Hanghang Liu) and D.L. All authors have read and agreed to the published version of the manuscript.

Funding: This research was funded by National Natural Science Foundation of China (Grant No. 51701225), the Project to Strengthen Industrial Development at the Grass-roots Level (Grant No. TC190A4DA/35), Innovation project of the cutting-edge basic research and key technology by SYNL (Grant No. L2019R36), Young Talent Project by SYNL (Grant No. L2019F33) and China Postdoctoral Science Foundation (Grant No. 2019M661153).

Conflicts of Interest: The authors declare no conflict of interest.

References

1. Medvedeva, A.; Bergström, J.; Gunnarsson, S.; Andersson, J. High-temperature properties and microstructural stability of hot-work tool steels. *Mater. Sci. Eng. A* **2009**, *523*, 39–46. [CrossRef]
2. Roberts, G.; Krauss, G.; Kennedy, R. *Tool Steels*, 5th ed.; ASM International: Novelty, OH, USA, 1998.
3. Jilg, A.; Seifert, T. Temperature dependent cyclic mechanical properties of a hot work steel after time and temperature dependent softening. *Mater. Sci. Eng. A* **2018**, *721*, 96–102. [CrossRef]
4. Chander, S.; Chawla, V. Failure of hot forging dies—An updated perspective. *Mater. Today Proc.* **2017**, *4*, 1147–1157. [CrossRef]
5. Pešička, J.; Kužel, R.; Dronhofer, A.; Eggeler, G. The evolution of dislocation density during heat treatment and creep of tempered martensite ferritic steels. *Acta Mater.* **2003**, *51*, 4847–4862. [CrossRef]
6. Zhang, Z.; Delagnes, D.; Bernhart, G. Microstructure evolution of hot-work tool steels during tempering and definition of a kinetic law based on hardness measurements. *Mater. Sci. Eng. A* **2004**, *380*, 222–230. [CrossRef]
7. Zhou, Q.; Wu, X.; Shi, N.; Li, J.; Min, N. Microstructure evolution and kinetic analysis of DM hot-work die steels during tempering. *Mater. Sci. Eng. A* **2011**, *528*, 5696–5700. [CrossRef]
8. Hu, X.; Li, L.; Wu, X.; Zhang, M. Coarsening behavior of M23C6 carbides after ageing or thermal fatigue in AISI H13 steel with niobium. *Int. J. Fatigue* **2006**, *28*, 175–182. [CrossRef]
9. Zhu, J.; Zhang, Z.; Xie, J. Improving strength and ductility of H13 die steel by pre-tempering treatment and its mechanism. *Mater. Sci. Eng. A* **2019**, *752*, 101–114. [CrossRef]
10. Malheiros, L.R.C.; Rodriguez, E.A.P.; Arlazarov, A. Mechanical behavior of tempered martensite: Characterization and modeling. *Mater. Sci. Eng. A* **2017**, *706*, 38–47. [CrossRef]
11. Michaud, P.; Delagnes, D.; Lamesle, P.; Mathon, M.H.; Levaillant, C. The effect of the addition of alloying elements on carbide precipitation and mechanical properties in 5% chromium martensitic steels. *Acta Mater.* **2007**, *55*, 4877–4889. [CrossRef]
12. Delagnes, D.; Lamesle, P.; Mathon, M.H.; Mebarki, N.; Levaillant, C. Influence of silicon content on the precipitation of secondary carbides and fatigue properties of a 5%Cr tempered martensitic steel. *Mater. Sci. Eng. A* **2005**, *394*, 435–444. [CrossRef]
13. Saha, D.C.; Biro, E.; Gerlich, A.P.; Zhou, Y. Effects of tempering mode on the structural changes of martensite. *Mater. Sci. Eng. A* **2016**, *673*, 467–475. [CrossRef]

14. Gomes, C.; Kaiser, A.-L.; Bas, J.-P.; Aissaoui, A.; Piette, M. Predicting the mechanical properties of a quenched and tempered steel thanks to a "tempering parameter". *Rev. Metall.* **2010**, *107*, 293–302. [CrossRef]

15. Revilla, C.; López, B.; Rodriguez-Ibabe, J.M. Carbide size refinement by controlling the heating rate during induction tempering in a low alloy steel. *Mater. Des.* **2014**, *62*, 296–304. [CrossRef]

16. Furuhara, T.; Kobayashi, K.; Maki, T. Control of cementite precipitation in lath martensite by rapid heating and tempering. *ISIJ Int.* **2004**, *44*, 1937–1944. [CrossRef]

17. Hernandez, V.H.B.; Nayak, S.S.; Zhou, Y. Tempering of martensite in dual-phase steels and its effects on softening behavior. *Metall. Mater. Trans. A* **2011**, *42*, 3115–3129. [CrossRef]

18. Nayak, S.S.; Hernandez, V.H.B.; Zhou, Y. Effect of chemistry on nonisothermal tempering and softening of dual-phase steels. *Metall. Mater. Trans. A* **2011**, *42*, 3242–3248. [CrossRef]

19. Bhadeshia, H.; Honeycombe, R. Tempering of Martensite. In *Steels: Microstructure and Properties*, 4th ed.; Gifford, C., Ed.; Butterworth-Heinemann: Oxford, UK, 2017; pp. 237–270.

20. Mote, V.D.; Purushotham, Y.; Dole, B.N. Williamson-Hall analysis in estimation of lattice strain in nanometer-sized ZnO particles. *J. Theoret. Appl. Phys.* **2012**, *6*, 6. [CrossRef]

21. Dini, G.; Ueji, R.; Najafizadeh, A.; Monir-Vaghefi, S.M. Flow stress analysis of TWIP steel via the XRD measurement of dislocation density. *Mater. Sci. Eng. A* **2010**, *527*, 2759–2763. [CrossRef]

22. Dragomir, I.C.; Li, D.S.; Castello-Branco, G.A.; Garmestani, H.; Snyder, R.L.; Ribarik, G.; Ungar, T. Evolution of dislocation density and character in hot rolled titanium determined by X-ray diffraction. *Mater. Charact.* **2005**, *55*, 66–74. [CrossRef]

23. ABaghdadi, H.; Rajabi, A.; Selamat, N.F.M.; Sajuri, Z.; Omar, M.Z. Effect of post-weld heat treatment on mechanical behavior and dislocation density of friction stir welded Al6061. *Mater. Sci. Eng. A* **2019**, *754*, 728–734. [CrossRef]

24. Liu, H.; Fu, P.; Liu, H.; Sun, C.; Sun, M.; Li, D. A novel large cross-section quenching and tempering mold steel matching excellent strength–hardness–toughness properties. *Mater. Sci. Eng. A* **2018**, *737*, 274–285. [CrossRef]

25. Chen, Y.-W.; Huang, B.-M.; Tsai, Y.-T.; Tsai, S.-P.; Chen, C.-Y.; Yang, J.-R. Microstructural evolutions of low carbon Nb/Mo-containing bainitic steels during high-temperature tempering. *Mater. Charact.* **2017**, *131*, 298–305. [CrossRef]

26. Jang, J.H.; Lee, C.-H.; Heo, Y.-U.; Suh, D.-W. Stability of (Ti, M)C (M= Nb, V, Mo and W) carbide in steels using first-principles calculations. *Acta Mater.* **2012**, *60*, 208–217. [CrossRef]

27. Wang, Z.; Zhang, H.; Guo, C.; Liu, W.; Yang, Z.; Sun, X.; Zhang, Z.; Jiang, F. Effect of molybdenum addition on the precipitation of carbides in the austenite matrix of titanium micro-alloyed steels. *J. Mater. Sci.* **2016**, *51*, 4996–5007. [CrossRef]

28. Chen, C.Y.; Yen, H.W.; Kao, F.H.; Li, W.C.; Huang, C.Y.; Yang, J.R.; Wang, S.H. Precipitation hardening of high-strength low-alloy steels by nanometer-sized carbides. *Mater. Sci. Eng. A* **2009**, *499*, 162–166. [CrossRef]

29. Krajnik, P.; Kopač, J. Modern machining of die and mold tools. *J. Mater. Process. Technol.* **2004**, *157–158*, 543–552. [CrossRef]

30. Bhadeshia, H.; Honeycombe, R. *Formation of Martensite, in Steels: Microstructure and Properties*, 4th ed.; Gifford, C., Ed.; Butterworth-Heinemann: Oxford, UK, 2017; pp. 135–177.

31. Sakamoto, H.; Otsuka, K.; Shimizu, K. Rubber-like behavior in a Cu-Al-Ni alloy. *Scr. Metall.* **1977**, *11*, 607–611. [CrossRef]

32. Kwon, H.; Lee, K.B.; Yang, H.R.; Lee, J.B.; Kim, Y.S. Secondary hardening and fracture behavior in alloy steels containing Mo, W, and Cr. *Metall. Mater. Trans. A* **1997**, *28*, 775–784. [CrossRef]

33. Williamson, G.K.; Smallman, R.E., III. Dislocation densities in some annealed and cold-worked metals from measurements on the X-ray debye-scherrer spectrum. *Philos. Mag.* **1956**, *1*, 34–46. [CrossRef]

34. Williamson, G.K.; Hall, W.H. X-ray line broadening from filed aluminum and wolfram. *Acta Metall.* **1953**, *1*, 22–31. [CrossRef]

35. Johnson, W.A.; Mehl, R.F. Reaction kinetics in processes of nucleation and growth. *Trans. Am. Inst. Min. Met. Eng.* **1939**, *135*, 416–442.

36. Avrami, M. Kinetics of phase change I. *J. Chem. Phys.* **1939**, *7*, 1103–1112. [CrossRef]

37. Avrami, M. Kinetics of phase change II. *J. Chem. Phys.* **1940**, *8*, 212–224. [CrossRef]

38. Avrami, M. Granulation, Phase change, and microstructure kinetics of phase change III. *J. Chem. Phys.* **1941**, *9*, 177–184. [CrossRef]

39. Watté, P.; van Humbeeck, J.; Aernoudt, E.; Lefever, I. Strain ageing in heavily drawn eutectoid steel wires. *Scripta Mater.* **1996**, *34*, 89–95. [CrossRef]

Article

Simulation and Prediction of the Vickers Hardness of AZ91 Magnesium Alloy Using Artificial Neural Network Model

Alaa F. Abd El-Rehim [1,2,*], **Heba Y. Zahran** [1,2], **Doaa M. Habashy** [2] and **Hana M. Al-Masoud** [1]

[1] Physics Department, Faculty of Science, King Khalid University, P.O. Box 9004, Abha 61413, Saudi Arabia; heldemardash@kku.edu.sa (H.Y.Z.); 435820327@kku.edu.sa (H.M.A.-M.)
[2] Physics Department, Faculty of Education, Ain Shams University, P.O. Box 5101, Heliopolis, Roxy, Cairo 11771, Egypt; doaamahmoud@edu.asu.edu.eg
* Correspondence: alaa.abdelrehim@kku.edu.sa

Received: 9 February 2020; Accepted: 8 April 2020; Published: 10 April 2020

Abstract: In this study, an artificial neural network (ANN) model was used to simulate and predict the Vickers hardness of AZ91 magnesium alloy. The samples of AZ91 alloy were aged at different temperatures (T_a = 100 to 300 °C) for different durations (t_a = 4 to 192 h) followed by water quenching at 25 °C. The age-hardening response of the samples was investigated by hardness measurements. The microstructure investigations showed that only discontinuous precipitates formed at low aging temperatures (100 and 150 °C), while continuous precipitates invaded all the samples at a high aging temperature (300 °C). Both discontinuous and continuous precipitates formed at the intermediate aging temperatures (200 and 250 °C). X-ray diffraction (XRD) analysis revealed that the microstructure comprised two phases: The α-Mg matrix and intermetallic β-Mg$_{17}$Al$_{12}$ phase. The alteration of the crystalline lattice parameters a, c, and c/a ratio with the aging time at various aging temperatures was also investigated. Both c and c/a ratio had the same behavior with aging time while a had an inverse trend. The observed variations of the lattice parameters were attributed to the mode of precipitation in AZ91 alloy. The ANN findings for the simulation and prediction perfectly conformed to the experimental data.

Keywords: AZ91 magnesium alloys; age-hardening response; microstructure evolution; β-Mg$_{17}$Al$_{12}$ phase; artificial neural network model

1. Introduction

Magnesium alloys play an important role in engineering applications, on account of their excellent properties such as high specific strength, high specific stiffness, good castability, excellent machinability, and abundant resources. These properties improve energy efficiency and decrease carbon dioxide emissions as well as other greenhouse gases [1,2]. Among various magnesium alloys, AZ91 alloy (Mg-9 wt.% Al-1 wt.% Zn) has superior resistance to corrosion and good mechanical characteristics [3]. It is well established [4] that the solid solubility of Al in α-Mg matrix attains its maximum value (12.7 wt.%) at 437 °C for the Mg-Al binary system. Zinc is added to this binary system to reduce the high solid solubility of Al during the aging process. As the Al/Zn ratio exceeds 3:1, only β-Mg$_{17}$Al$_{12}$ phase precipitates. Consequently, corrosion resistance and mechanical properties will be improved [2].

The β-Mg$_{17}$Al$_{12}$ phase can be precipitated in the AZ91 alloys continuously or discontinuously. The lamellar structure of discontinuous precipitation (DP) forms along the grain boundaries of α-Mg [5–10]. The continuous precipitation (CP) occurs in the form of lath-type or fine plate-type morphologies inside α-Mg grains which have not been occupied by DP [11–14]. At low aging temperatures, DP is favored while CP invades all samples at high aging temperatures. The continuous, as well as the discontinuous precipitates, are formed and controlled by the aging treatment conditions. They can simultaneously or competitively appear at the aging temperatures ranging from 175 to 270 °C [15–17].

The mechanical characteristics of the as-cast Mg alloys, especially AZ91 alloy, can be improved through the precipitation hardening during the aging treatment process. Several research reports [18–22] have documented the variations in the mechanical characteristics of different Mg-based alloys with the aging treatment. Kim et al. [2] concluded that the changes in the mechanical characteristics during the aging process are associated with the alteration of β-$Mg_{17}Al_{12}$ precipitation behavior. The effect of precipitation of β-$Mg_{17}Al_{12}$ phase on the creep behavior of AZ91 alloy aged at 150 and 200 °C has also been studied [7]. The results showed that the β-$Mg_{17}Al_{12}$ phase suffers severe cracking, leading to early failure. Robson and Paa-Rai [18] showed that insoluble Zn-Zr particles were present in the microstructure of Mg-6Zn-0.6Zr (wt.%) alloy even after solution heat treatment. Therefore, in Zn-Zr, particles most likely remained intact after homogenization treatment. Li et al. [19] concluded that the Mg-5Zn-2Gd-0.4Zr (wt.%) alloy showed a significant age-hardening response with a hardness increment from 62 HV to 72 HV by aging at 200 °C for up to 80 h. Xia et al. [20] investigated the age-hardening behavior of Mg-4Sm-xZn-0.4Zn (x = 0, 0.3, 0.6, and 1.3 wt.%) alloy. They reported that all samples exhibited significant precipitation hardening effect. The optimum composition was determined as Mg-4Sm-(0.3-0.6)Zn-0.4Zr alloy. Liu et al. [22] studied the microstructure and precipitation behavior of Mg-4Y-2Zn (at.%) extruded alloy during solution treatment and aging processes. During the aging process, the nanoscale β' phases were coherent with α-Mg precipitated in the matrix. Suresh et al. [23] stated that the addition of 0.2 wt.% charcoal to the AZ91 alloy can significantly refine the microstructure and accelerate the aging kinetics of refined alloys.

An artificial neural network (ANN) model is commonly used for both simulation and prediction of the mechanical properties of metals and alloys. The ANN is inspired by a biological neural network, consisting of interconnected artificial nodes or neurons that can model complicated functional relationships [24]. Okuyucu et al. [25] predicted yield strength, elongation, and tensile strength of aluminum friction stir welding joints at heat affected zone (HAZ) with mean error 0.656, 7.596, and 1.650, respectively, whereas Asadi et al. [26] anticipated the hardness profile and grain size of AZ91/SiC of friction stir processing (FSP) nanocomposite plate with maximum training error 1.8% and 0.5%, respectively. Yousif et al. [27] established an ANN model for the prediction of tensile stress, bending stress, and elongation of friction stir welding of AA6061 aluminum. The errors of tensile stress, bending stress, and elongation were reported to be 1.7524%, 7.3777%, and 11.98%, respectively. Ghetiya et al. [28] used ANN with 4–8–1 architecture to predict the tensile strength of friction stir welding (FSW) joint with less than 3% error. In their research, Arunchai et al. [29] employed the ANN to model the resistance spot welding (RSW) joints with an accuracy of 95%. Ansari et al. [24] presented a computational model based on ANN to analyze the friction stir extrusion (FSE) process of magnesium. Multilayer neural network was used to discover the correlation between FSE parameters and average grain size of the produced wires. The accuracy of the developed model can be shown through root mean square error (RMSE) and linear regression analyses. Recently, Habashy et al. [30] applied the ANN for modeling the composite hardness (H_c), yield stress (σ_y), and film hardness (H_f) of titanium dioxide nanoparticles at different temperatures, dwell times, and relative indentation depths (β). Mean squared error values for the prediction of H_c, σ_y, and H_f were found to be 1.4369×10^{-16}, 3.9368×10^{-16}, and 6.807×10^{-18}, respectively.

The present work intended to investigate the applicability of ANN model to simulate and predict the age-hardening response of AZ91 alloy. The ANN results were compared with the measured experimental data.

2. Experimental Procedures

The AZ91 magnesium alloy chemical composition obtained by Magnesium Elektron is listed in Table 1. Block samples with the dimensions 15 mm × 15 mm × 10 mm of the studied alloy were designed for hardness measurements and microstructure characterization. In order to achieve solution annealing, the specimens were exposed to solution heat treatment at a temperature of 420 °C for 24 h in a protective argon atmosphere. The annealing furnace (model no. T-5.0x10-w-200, Heraeus type, Germany), power 380 V/50 Hz, 80 Amp was used in the present work. Temperature variations during heat treatment were recorded by a 0.8-mm diameter chromel-alumel (type K) thermocoupler, which was connected to a computer-based acquisition system. Later, all samples were dropped into the water at 25 °C (room temperature) so as to quench and conserve the supersaturated solid solution. Finally, the solution-treated specimens were aged at various temperatures (T_a = 100, 150, 200, 250, and 300 °C) for different durations (t_a = 4, 8, 12, 24, 48, 96, 144, and 192 h) followed by water quenching at 25 °C in order to study the age-hardening response of AZ91 alloy as shown in Figure 1. The temperature measurement accuracy was ±1 °C.

Table 1. Chemical compositions (wt.%) of the alloy used in the present study.

Al	Zn	Mn	Si	Cu	Fe	Mg
8.4	0.27	0.09	<0.02	<0.001	<0.001	Bal.

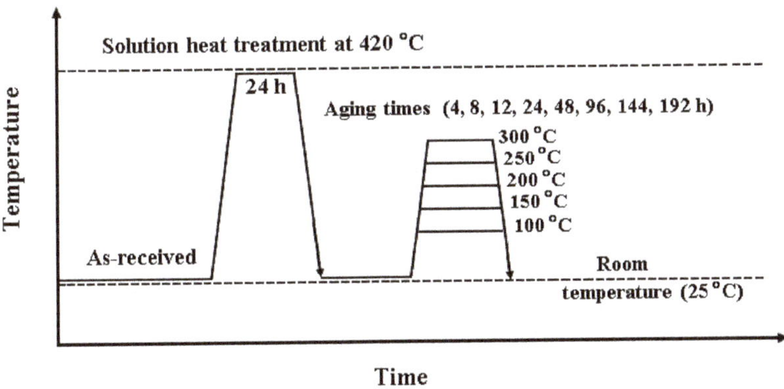

Figure 1. A schematic diagram showing the heat treatment procedure.

To reveal the microstructure of the investigated samples, they were ground and polished according to the usual magnesium alloys' procedures [31]. Then, the prepared specimens were etched for about 90 s in a nitric acid solution (4 mL HNO3 and 96 mL C2H5OH). Afterward, the samples were cleaned with anhydrous ethyl alcohol and then dried with air blast. The microstructure examinations were performed on a light microscope Olympus BH2 (Olympus Co., Ltd., Bangkok, Thailand) equipped with a Leica DC 200 MTV-3 camera. A Shimadzu D6000 X-ray diffractometer (Shimadzu Corporation, Tokyo, Japan) with Cu-Kα radiation operated at 30 kV and 30 mA (with wavelength λ of 0.15406 nm) was used to distinguish the formed phases.

The age-hardening response was investigated using a Vickers hardness testing machine under a load of 0.5 kg for 10 s at room temperature. To ensure reproducibility, the average of 10 random indentations was taken to calculate each reported hardness value.

3. Artificial Neural Network (ANN)

A nonlinear function mapping of the input variables in the corresponding network output variables was provided by the artificial neural network (ANN), and it was not required to have the real mathematics form of the relation between the input and output variables [30]. The most commonly used type of feedforward neural network is multilayer perceptron (MLP). MLP neural network has many layers of simple neurons which are coordinated in a manner that every neuron in a layer is linked to every neuron in the following by weight (see Figure 2). These layers are an input layer, at least one hidden layer, and an output layer. The sum of weighted inputs was calculated by all neuron and then converted by the next transfer function:

$$n_j = \frac{1}{1 + \exp(-x)}$$

where n_j is the output of the j-th neuron, and x is provided with:

$$x = \sum_{i=1}^{n} w_{ij} p_i + b_j$$

where w_{ij} is the weights applied from the i-th neurons in the previous layer to the j-th neurons, p_i is the output from the i-th neuron, and b is a bias term.

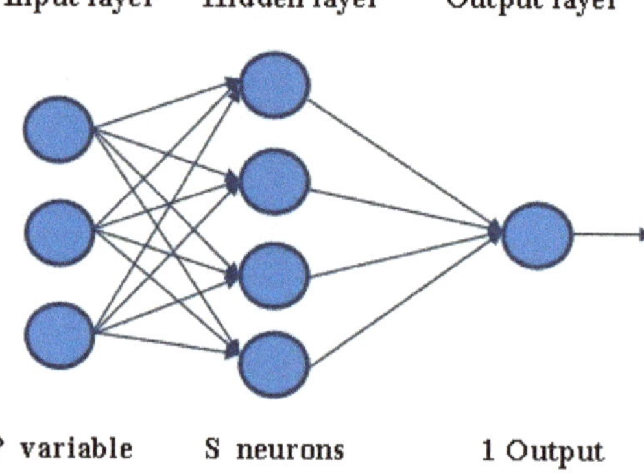

Figure 2. Topology of conventional P–S–1 MLP.

The MLP networks utilize a supervised learning technique. Both back propagation error (BPE) and training data, in the training algorithm, are provided to the networks by the weights adjustment and biases until the expected values for the network are consistent with the actual values. This adjustment is made with the BPE training algorithm by the comparison between the actual value t_{ij} and the expected values a_{ij} of the network by calculating the total sum of the square error (SSE) for the n data of the training dataset,

$$SSE = \sum_{i=1}^{n} \left(t_{ij} - a_{ij}\right)^2$$

In this study, the ANN model based on multilayer perceptron was used to compute Vickers hardness values. The ANN model was configured to have aging temperatures and aging times as inputs and Vickers hardness values as output, as shown in Figure 3. Various network settings were attempted to provide the best mean square errors (MSE) and the best results using the input-output scheme. The three hidden layers' configuration with different neurons were chosen. In every hidden layer, the number of neurons was 34, 24, and 25, respectively. The transfer functions were logsig function for all hidden layers and linear pureline function for output layer.

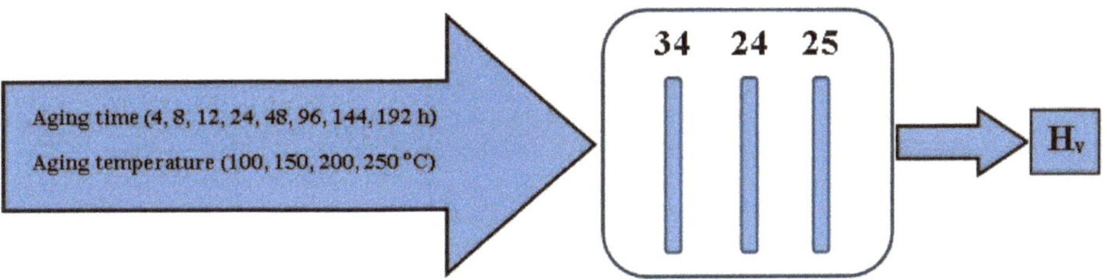

Figure 3. Block diagram of ANN based modeling.

4. Results and Discussion

Figure 4 depicted the age-hardening response of AZ91 magnesium alloy at various aging temperatures. The dashed lines interpolate the experimental points represented by the markers. A little augment in the hardness with increasing aging time was detected for specimens aged at 100 °C. The hardness values started to increase considerably when aging at 150 °C without the attainment of its peak hardness within the examined time interval. After aging at 200 °C, the hardness considerably improved with the extension of aging time, achieving its peak value after 24 hours, and, subsequently, further aging times reduced the hardness values. The same tendency was found at the 250 °C aging curve as that at 200 °C, but with the lower hardness values. For 250 °C aging curve, the peak hardness was reached after about 12 h, which was 12 h earlier than that at 200 °C. This points out that the age-hardening response of AZ91 was accelerated at a higher temperature. No major hardness changes were observed for the AZ91 alloy aged at 300 °C.

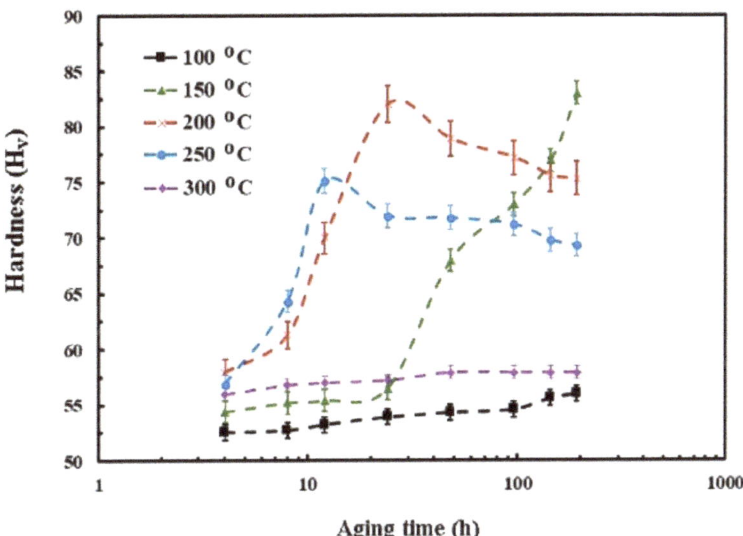

Figure 4. Age-hardening curves of AZ91 magnesium alloy aged at different temperatures.

The vital agents that dominate the characteristics of precipitation-hardened alloys are the size, type, distribution, and morphology of the new strengthening precipitates in the original matrix [32]. The variations in the age-hardening curves at various aging temperatures can be ascribed to the interaction between the moving dislocations and β-$Mg_{17}Al_{12}$ precipitates. At the lower aging temperature of 100 °C, the hardness increased slightly with an increase in aging time from 4 to 192 h. The microstructure of AZ91 samples aged at 100 °C for 4 and 192 h are, respectively, shown in

Figure 5a,b. The microstructure investigation showed the existence of α-Mg as a matrix and there was no β-$Mg_{17}Al_{12}$ phase for sample aged for 4 h (Figure 5a). This may be rendered to that 4 hours were insufficient aging time to produce the DP. With the increment in the aging duration up to 192 h, the discontinuous precipitation colonies of the β ($Mg_{17}Al_{12}$) phase (dark contrast of the second phase) started to form within the primary α-Mg matrix (light background) (Figure 5b). The existence of a slight volume fraction of the second phase (β-$Mg_{17}Al_{12}$) would lead to a little increase in the hardness.

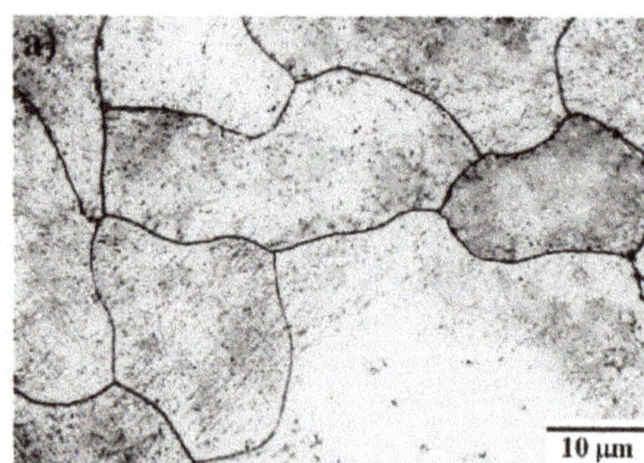
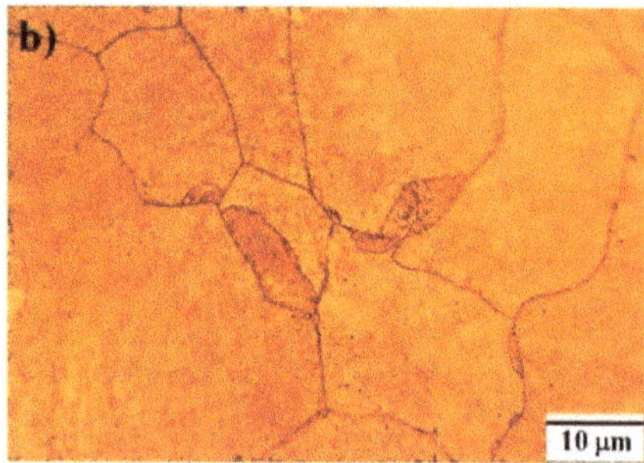

Figure 5. Optical micrographs of AZ91 alloy aged at 100 °C for (**a**) 4 h and (**b**) 192 h.

When the samples aged at 150 °C, the hardness values improved continuously with the aging time increment. No peak hardness was detected for AZ91 samples aged at 150 °C even after aging for 192 h. Our results are consistent with those reported by Celotto [33], who found that the peak hardness at 150 °C reached 100 HV after aging for 10,000 h. Figure 6a–d displays typical optical images of samples aged at 150 °C for 4, 24, 96, and 192 h, respectively. It is obviously seen that the microstructure containing the β ($Mg_{17}Al_{12}$) phase existed at the α-Mg grain boundaries. By extending aging time, the nucleation sites of DP increased and the previously formed DP continued to expand towards the inside grains and into the adjacent grains as distinguished from the microstructures of samples aged for 24 and 96 h (Figure 6b,c). After aging for 192 h, the sample was invaded by the DP (Figure 6d). It has been reported [34] that the β ($Mg_{17}Al_{12}$) phase nucleated at the grain boundaries then developed as the grain boundaries, migrating towards the adjacent grains. The depletion of Al solute atoms behind grain boundaries should associate this migration. Zheng et al. [35] indicated that the migration of the grain boundary would enhance the nucleation and growth of DP. Therefore, the grain boundary migration performed a significant role in the nucleation and growth of DP [36]. Figure 6 confirms that the volume fraction of DPs increased with the aging time, resulting in higher hardness values (Figure 4).

Figure 7 exhibits typical optical images of samples aged at 200 °C for different times. After the aging treatment for 4 h, it was detected the lamellar structure of the DPs distributed along the grain boundaries of the α-Mg matrix (Figure 7a). The microstructure evolution showed that the whole sample had been invaded by the DP as the aging time increased from 4 to 24 hours (Figure 7b). However, when the sample aged for 24 hours, the maximum hardness value was achieved. The DPs that acted as an impediment to dislocation motion led to a maximum hardness value. The reduction in the hardness values with the aging period interval (24–92 h) might be attributed to the existence of both CPs and DPs (Figure 7c). It has been stated [37,38] that the growth rate of discontinuous precipitates would terminate if the grains were filled with the DPs or if CPs began to grow considerably and hinder the propagation of the discontinuous precipitates. Thus, the volume fraction of continuous precipitates grew as the aging time increased from 24 to 96 h, and that of discontinuous precipitation gradually reduced, resulting in lower hardness values. Earlier studies [39,40] declared that discontinuous precipitates were effectively inhibited as the continuous precipitates were formed. Dissimilar to the microstructure of the sample aged for 92 h (Figure 7c), plenty of lath-shaped precipitates occupied the whole sample, and no discontinuous precipitates were detected (Figure 7d). Since the CPs consisting of coarsely lath-shaped precipitates lie on the base plane of matrix, there are great opportunities for dislocations to bypass the obstacles leading to lower hardness values. Comparing the micrographs shown in Figures 7 and 8, it is clear that the microstructure is identical, while the volume fraction of β-phase differs. The samples aged at 200 °C had volume fraction of β-phase that was considerably denser than that of the 250 °C samples, which could explain the higher hardness values for samples aged at 200 °C. The hardness response for all the AZ91 samples aged at 300 °C for various times (4–192 h) was nearly constant, and it was found to be lower than that of the samples aged at 250 °C. The low hardness response for aging at 300 °C may be rendered to the formation, growth, and coarsening of CPs. The CPs of the small number and large size were less effective to prevent the dislocations' motion. This trend would lead to a loss of hardness. The CPs' precipitating during aging at 300 °C in the present work agrees with the previous results of AZ91 alloy investigations [8,33,41].

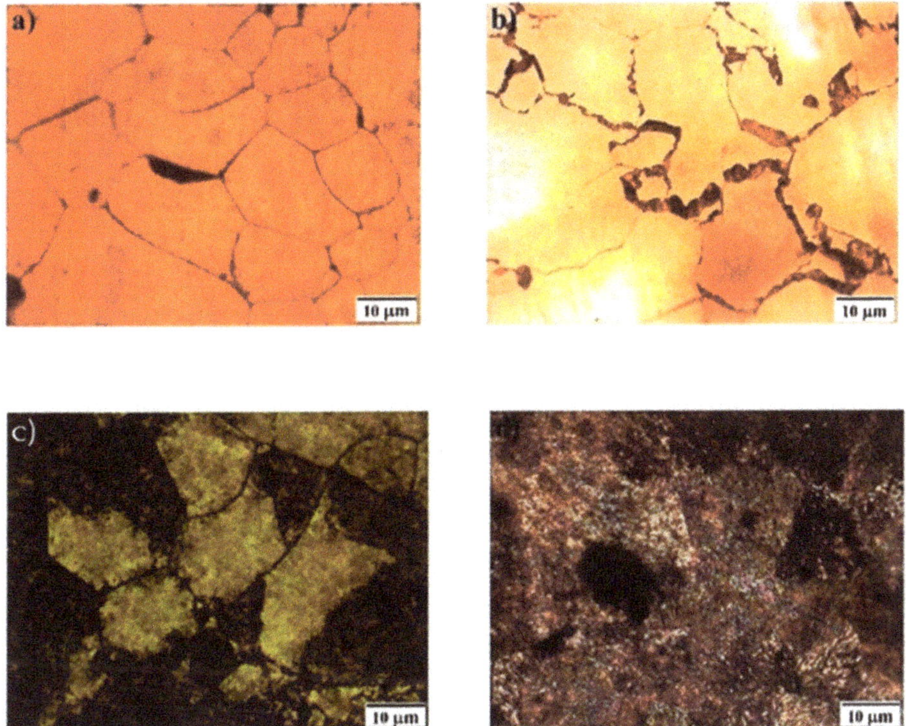

Figure 6. Optical micrographs of AZ91 alloy aged at 150 °C for (**a**) 4 h, (**b**) 24 h, (**c**) 96 h, and (**d**) 192 h.

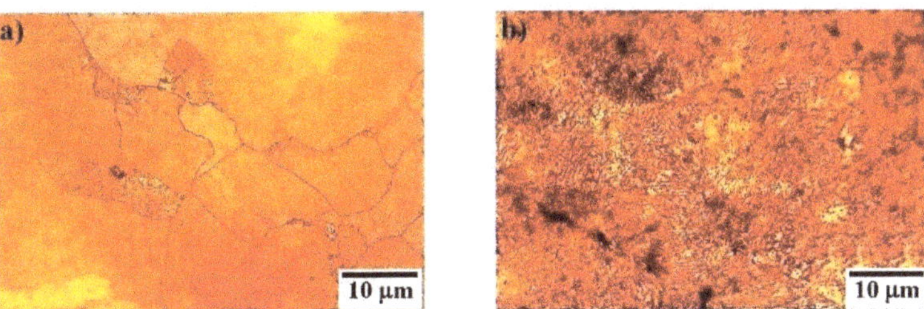

Figure 7. *Cont.*

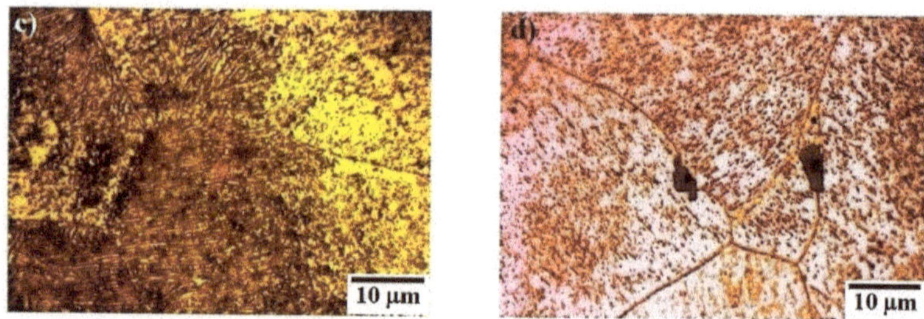

Figure 7. Optical micrographs of AZ91 alloy aged at 200 °C for (**a**) 4 h, (**b**) 24 h, (**c**) 96 h, and (**d**) 192 h.

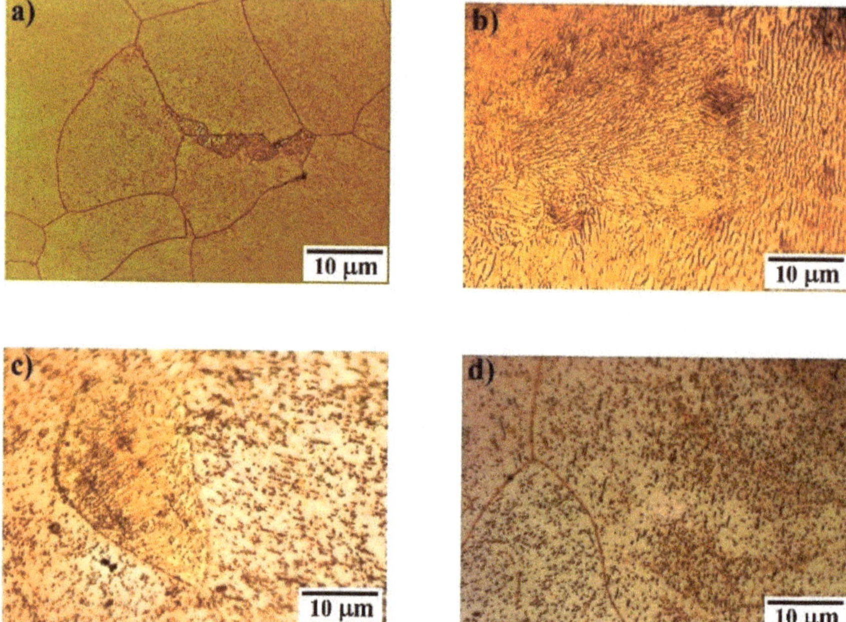

Figure 8. Optical micrographs of AZ91 alloy aged at 250 °C for (**a**) 4 h, (**b**) 12 h, (**c**) 96 h, and (**d**) 192 h.

X-ray diffraction (XRD) measurements were applied for the studied samples aged at different temperatures for various aging durations. The recorded XRD series for all the investigated samples included both α-Mg with hcp crystal structure according to JCPDS card no. 65-3365 and body-centered cubic crystal structured (β-Mg$_{17}$Al$_{12}$ phase) according to JCPDS card no. 73-1148. The crystalline lattice parameters a, c, and c/a ratio were calculated for α-Mg matrix at different aging temperatures and indexed with errors in Table 2 by the aid of X-ray data. The errors were calculated for a, c, and c/a ratio with respect to their standard values recorded in JCPDS card no. 65-3365 for α-Mg phase ($a = 3.208$ Å, $c = 5.21$ Å, and $c/a = 1.62406$). For α-Mg phase of hcp crystal structure, the interplanar distance, d, between these two planes (100) and (101) can be correlated to a and c lattice parameters through the following equation:

$$\frac{1}{d^2} = \frac{4}{3}\left(\frac{h^2 + hk + k^2}{a^2}\right) + \frac{l^2}{c^2} \tag{1}$$

Table 2. The calculated crystalline lattice parameters *a*, *c*, and *c/a* ratio for α-Mg matrix at different aging temperatures and various aging times.

Lattice Parameter	Aging Time (h)	Aging Temperature (°C)									
		100	Error %	150	Error %	200	Error %	250	Error %	300	Error %
a (Å)	4	3.175	1.03	3.179	0.9	3.194	0.43	3.19	0.56	3.183	0.79
	8	3.176	0.99	3.18	0.87	3.197	0.34	3.2	0.25	3.184	0.75
	12	3.177	0.96	3.181	0.85	3.21	0.06	3.23	0.69	3.185	0.72
	24	3.179	0.9	3.183	0.78	3.24	0.99	3.216	0.24	3.186	0.69
	48	3.181	0.84	3.204	0.12	3.227	0.58	3.21	0.062	3.187	0.65
	96	3.182	0.81	3.212	0.125	3.22	0.37	3.207	0.0312	3.188	0.62
	144	3.183	0.78	3.224	0.499	3.22	0.37	3.205	0.0935	3.188	0.62
	192	3.184	0.75	3.229	0.655	3.22	0.37	3.205	0.0935	3.188	0.62
c (Å)	4	5.213	0.058	5.21	0	5.19	0.384	5.21	0.192	5.213	0.115
	8	5.211	0.019	5.205	0.084	5.1736	0.699	5.206	0.787	5.211	0.134
	12	5.21	0	5.206	0.077	5.17	0.768	5.206	0.576	5.21	0.154
	24	5.209	0.019	5.202	0.077	5.185	0.479	5.202	0.465	5.209	0.249
	48	5.208	0.038	5.198	0.154	5.196	0.269	5.198	0.384	5.208	0.269
	96	5.207	0.058	5.179	0.595	5.177	0.633	5.179	0.578	5.207	0.288
	144	5.206	0.077	5.168	0.806	5.17	0.768	5.168	0.595	5.206	0.326
	192	5.205	0.096	5.158	0.998	5.168	0.806	5.158	0.614	5.205	0.326
c/a	4	1.642	1.098	1.639	0.912	1.625	0.053	1.630	0.371	1.635	0.683
	8	1.641	1.027	1.637	0.796	1.618	0.357	1.615	0.539	1.634	0.618
	12	1.640	0.976	1.636	0.778	1.611	0.829	1.604	1.253	1.633	0.567
	24	1.639	0.893	1.634	0.631	1.600	1.463	1.613	0.703	1.631	0.439
	48	1.637	0.81	1.622	0.106	1.610	0.853	1.617	0.446	1.630	0.388
	96	1.636	0.759	1.612	0.719	1.608	1.004	1.6151	0.547	1.629	0.338
	144	1.635	0.708	1.603	1.298	1.606	1.138	1.616	0.502	1.628	0.299
	192	1.634	0.657	1.597	1.642	1.605	1.176	1.6156	0.521	1.628	0.299

Figure 9a illustrates the variation of *a* with aging time for all the investigated samples aged at various temperatures. The lines joining the markers are guides for the eye. It is clearly seen that the trend of the lattice parameter, *a*, was in contrary to the behavior of the *c/a* ratio (Figure 9c) for the samples aged at 150, 200, and 250 °C. On the other hand, their values were approximately constant with the variation of the aging time at 100 and 300 °C. It is observed from Figure 9a, the crystalline lattice parameter, *a*, increased continuously with aging time at 150 °C. The values of *a* varied from 3.179 Å to 3.229 Å at the maximum aging time of 192 h. This variation equaled about 1.57% from its initial value at the minimum aging time of 4 h. At the aging temperature of 200 °C, the *a* values increased with increasing aging time up to 24 h, and then they decreased and remained constant. A similar trend was observed for the lattice parameter, *a*, at 250 °C but the reduction occurred after the lower aging time of 12 h. Figure 9b clarifies the variations of the lattice parameter, *c*, with aging time at different aging temperatures. From this figure, it is obviously seen that the parameter *c* had a value of about 5.211 Å, and still remained constant at 100 °C. Moreover, it had a value of about 5.204 Å at 300 °C. The lattice parameter, *c*, showed the same behavior of *c/a* ratio with aging time and temperature. These calculated values of *a* and *c* are matched with recorded values in JCPDS card no. 65-3365 for Mg matrix, and they agreed well with the values reported in the literature [39,41]. The separation of aluminum to form β-Mg$_{17}$Al$_{12}$ phase may be held responsible for the alteration in the lattice parameter values. Moreover, the precipitation mechanism was improved at 200 °C, which strongly affected the hardness values of the investigated alloy.

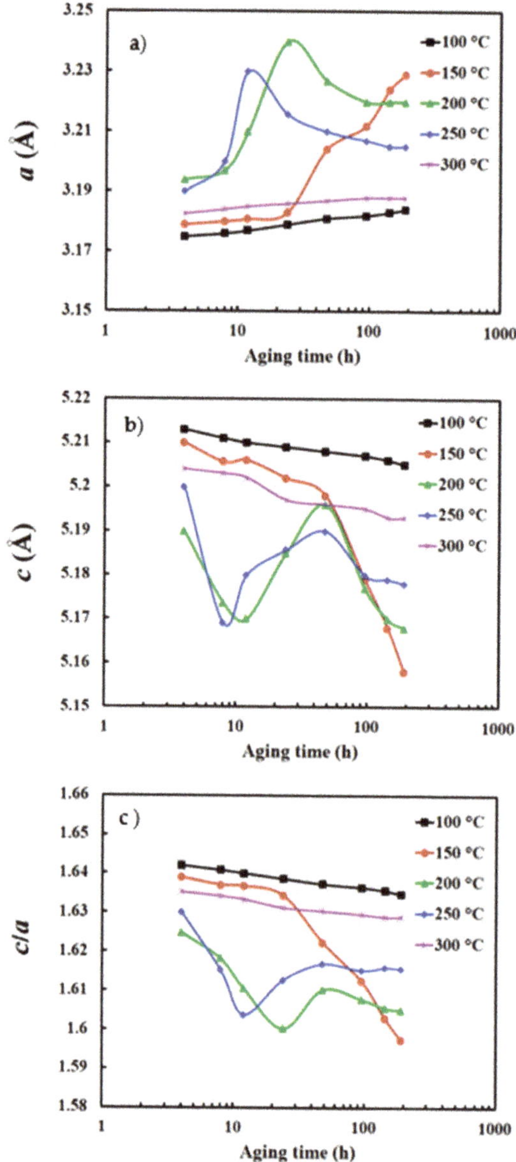

Figure 9. The variation of (**a**) the lattice parameter a, (**b**) the lattice parameter c, and (**c**) c/a ratio with the aging time for all the investigated samples aged at different temperatures.

The ANN model was used to simulate and predict the Vickers hardness of AZ91 magnesium alloy. Two inputs parameters, aging time (t_a) and aging temperature (T_a), were taken into account to simulate the Vickers hardness as an output parameter. The data were taken from the experimental results. The transfer function was selected as a logsig and pureline for hidden layers and output layer, respectively. Experimental data were used for training the model. The ANN was trained with measured datasets at 100, 150, 200, and 250 °C. The training was carried out for 541 iterations (epochs). The artificial neural network was endeavoring to get a better mean square error (MSE) and best execution for the network. The goal of the training was 10^{-5} (MSE = 10^{-5}) and we reached the nearest value (best) to this goal.

The result of training is shown in Figure 10. It is clearly seen that the mean squared error of the network decreased from a large value to a smaller value. Moreover, the network was learning. After the network memorized the training set, training was completed. The obtained equation (which represents the hardness, H_v) for network is represented in Appendix A. The best ANN model training was obtained according to the value of MSE (9.8256×10^{-6}) which represented the difference between simulation results and experimental data. The performance of the ANN model was validated by comparing the prediction values at 300 °C with the measured experimental data. The predicted values were in good agreement with the measured dataset at 300 °C. Results of simulation, prediction, and experimental values are shown in Figure 11. The age-hardening response was simulated at temperatures of 100, 150, 200, and 250 °C and predicted at temperature of 300 °C. Figure 11 depicts that the simulation and prediction results of the ANN were in good agreement with experimental data of the age-hardening response of the investigated alloy. Comparison between the ANN results and experimental data is depicted in Table 3. In summary, the present study showed that the ANN model can effectively simulate and predict the age-hardening response of AZ91 magnesium alloy.

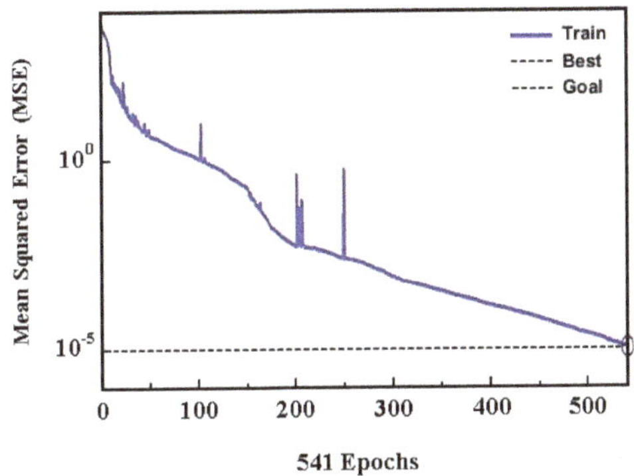

Figure 10. The ANN performance for hardness profile of AZ91 alloy.

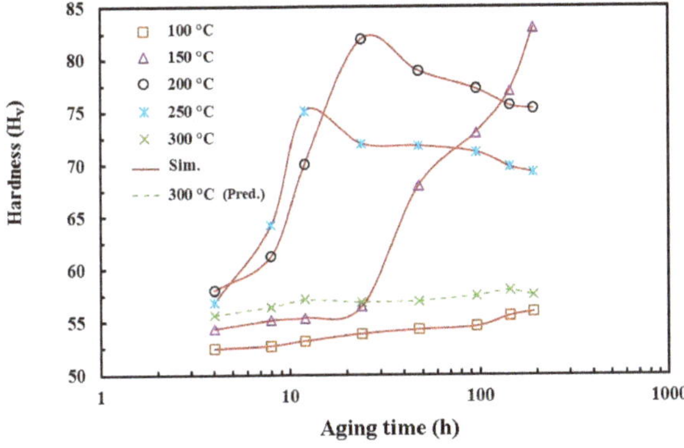

Figure 11. Simulation and prediction of the hardness profile based ANN model.

Table 3. Comparison between the ANN results and experimental data.

Aging Time (h)	Aging Temperature (°C)									
	100		150		200		250		300	
	Exp.	Sim.	Exp.	Sim.	Exp.	Sim.	Exp.	Sim.	Exp.	Pred.
4	52.50	52.48	54.40	54.40	58.00	58.00	56.90	56.88	56.00	55.70
8	52.70	52.74	55.21	55.19	61.21	61.30	64.30	64.25	56.80	56.40
12	53.20	53.18	55.40	55.40	70.00	70.00	75.12	75.10	57.00	57.15
24	53.89	53.89	56.50	56.50	82.00	81.95	72.00	72.00	57.20	56.86
48	54.30	54.29	68.11	68.00	78.90	78.89	71.80	71.80	57.90	56.93
96	54.60	54.60	73.00	73.00	77.23	77.20	71.25	71.21	57.90	57.50
144	55.60	55.59	77.00	76.90	75.58	75.60	69.80	69.80	57.90	58.01
192	56.00	56.00	83.00	82.90	75.33	75.30	69.30	69.30	57.90	57.56

5. Conclusions

In this article, the artificial neural network (ANN) model was adopted to simulate and predict the hardness profiles of AZ91 magnesium alloy. The results of ANN were compared with the experimental data and the following conclusions could be drawn from this study:

(1) The microstructure evolutions revealed that the mode of precipitation in AZ91 alloy was strongly affected by aging temperature and time of aging. The discontinuous precipitation was favored at low temperatures of aging (100 and 150 °C), while at high aging temperature (300 °C), continuous precipitation (CP) invaded all the samples. Both discontinuous and continuous precipitation reactions occurred at the intermediate aging temperature range (200–250 °C).

(2) The variations in the age-hardening response of AZ91 alloy at various aging temperatures was clarified in view of the size, morphology, and distribution of the β-$Mg_{17}Al_{12}$ precipitates.

(3) Based on the artificial neural network (ANN) model, simulated analysis showed a high correspondence with very low mean square error (MSE). The simulation results indicated higher precision of the operating model to the experimental data.

(4) The prediction results of the ANN model were found to be in good agreement with the measured dataset.

(5) The ANN model effectively simulated and predicted the age-hardening response of AZ91 magnesium alloy.

Author Contributions: Conceptualization, A.F.A.E.-R., H.Y.Z., and D.M.H.; investigation, A.F.A.E.-R., H.Y.Z., and H.M.A.-M.; methodology, A.F.A.E.-R., H.Y.Z., D.M.H., and H.M.A.-M.; writing—original draft, D.M.H. and H.M.A.-M.; writing—review and editing, A.F.A.E.-R. and H.Y.Z. All authors have read and agreed to the published version of the manuscript.

Funding: This research was funded by King Khalid University through General Research Project under grant number G.R.P/259/40.

Acknowledgments: The authors extend their appreciation to the Deanship of Scientific Research at King Khalid University for funding this work through General Research Project under grant number G.R.P/259/40.

Conflicts of Interest: The authors declare no conflict of interest.

Appendix A

The equation which describes hardness is given by:

H = pure line [net.LW{4, 3} logsig (net.LW{3, 2} logsig (net.LW{2, 1} logsig (net.IW{1, 1}T + net.b{1})+ net.b{2}) + net.b{3} + net.b{4}]

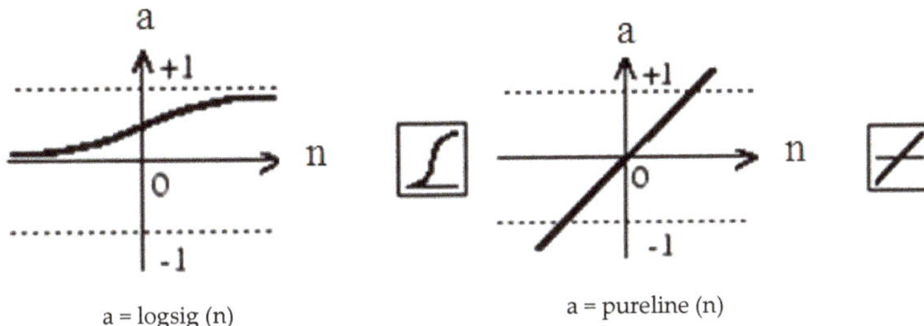

Where T is the inputs (temperatures and times), net.IW {1, 1} is linked weights between the input layer and first hidden layer, net.LW {2, 1} is linked weights between first and second hidden layer, net.LW {3, 2} is linked weights between the second and third layer, net.LW {4, 3} is linked weights between the third and output layer, net.b{1} is the bias of the first hidden layer, net.b {2} is the bias of the second hidden layer, net.b{3} is the bias of the third layer, and net.b {4} is the bias of the output layer.

References

1. Xiao, P.; Gao, Y.; Yang, X.; Xu, F.; Yang, C.; Li, B.; Li, Y.; Liu, Z.; Zheng, Q. Processing, microstructure and ageing behavior of in-situ submicronTiB2 particles reinforced AZ91 Mg matrix composites. *J. Alloys Compd.* **2018**, *764*, 96–106. [CrossRef]
2. Kim, S.H.; Lee, J.U.; Kim, Y.J.; Bae, J.H.; You, B.S.; Park, S.H. Accelerated precipitation behavior of cast Mg-Al-Zn alloy by grain refinement. *J. Mater. Sci. Technol.* **2018**, *34*, 265–276. [CrossRef]
3. Abd El-Rehim, A.F.; Zahran, H.Y.; Al-Masoud, H.M.; Habashy, D.M. Microhardness and microstructure characteristics of AZ91magnesium alloy under different cooling rate conditions. *Mater. Res. Express* **2019**, *6*, 086572. [CrossRef]
4. Nie, J.F. Precipitation and hardening in magnesium alloys. *Metall. Mater. Trans. A* **2012**, *43*, 3891–3939. [CrossRef]
5. Lee, J.U.; Kim, S.H.; Kim, Y.J.; Park, S.H. Effects of homogenization time on aging behavior and mechanical properties of AZ91 alloy. *Mater. Sci. Eng. A* **2018**, *714*, 49–58. [CrossRef]
6. Jun, J.H. Dependence of hardness on interlamellar spacing of discontinuous precipitates in cast AZ91 magnesium alloy. *J. Alloys Compd.* **2017**, *725*, 237–241. [CrossRef]
7. Srinivasan, A.; Ajithkumar, K.K.; Swaminathan, J.; Pillai, U.T.S.; Pai, B.C. Creep behavior of AZ91 magnesium alloy. *Procedia Eng.* **2013**, *55*, 109–113. [CrossRef]
8. Braszczynska-Malik, K.N. Discontinuous and continuous precipitation in magnesium-aluminium type alloys. *J. Alloys Compd.* **2009**, *477*, 870–876. [CrossRef]
9. Yu, S.; Gao, Y.; Liu, C.; Han, X. Effect of aging temperature on precipitation behavior and mechanical properties of extruded AZ80-Ag alloy. *J. Alloys Compd.* **2015**, *646*, 431–436. [CrossRef]
10. Lai, W.J.; Li, Y.Y.; Hsu, Y.F.; Trong, S.; Wang, W.H. Aging behaviour and precipitate morphologies in Mg-7.7Al-0.5Zn-0.3Mn (wt.%) alloy. *J. Alloys Compd.* **2009**, *476*, 118–124. [CrossRef]
11. Zhou, J.P.; Zhao, D.S.; Wang, R.H.; Sun, Z.F.; Wang, J.B.; Gui, J.N.; Zheng, O. In Situ observation of ageing process and new morphologies of continuous precipitates in AZ91 magnesium alloy. *Mater. Lett.* **2007**, *61*, 4707–4710. [CrossRef]
12. Zhang, M.X.; Kelly, P.M. Crystallography of $Mg_{17}Al_{12}$ precipitates in AZ91D alloy. *Scripta Materialia* **2003**, *48*, 647–652. [CrossRef]
13. Duly, D.; Simon, J.P.; Brechet, Y. On the competition between continuous and discontinuous precipitations in binary Mg-Al alloys. *Acta Metallurgica Materialia* **1995**, *43*, 101–106.
14. Sahoo, B.N.; Panigrahi, S.K. Effect of In Situ (TiC-TiB2) reinforcement on aging and mechanical behavior of AZ91 magnesium matrix composite. *Mater. Charact.* **2018**, *139*, 221–232. [CrossRef]
15. Celotto, S.; Bastow, T.J. Study of precipitation in aged binary Mg–Al and ternary Mg–Al–Zn alloys using 27Al NMR spectroscopy. *Acta Materialia* **2001**, *49*, 41–51. [CrossRef]
16. Bastow, T.J.; Celotto, S. Clustering and formation of nano-precipitates in dilute aluminium and magnesium alloys. *Mater. Sci. Eng. C* **2003**, *23*, 757–762. [CrossRef]

17. Xiuqing, Z.; Lihua, L.; Naiheng, M.; Haowei, W. Effect of aging hardening on In Situ synthesis magnesium matrix composites. *Mater. Chem. Phys.* **2006**, *96*, 9–15. [CrossRef]

18. Robson, J.D.; Paa-Rai, C. The interaction of grain refinement and ageing in magnesium–zinc–zirconium (ZK) alloys. *Acta Materialia* **2015**, *95*, 10–19. [CrossRef]

19. Li, J.H.; Barrirero, J.; Sha, G.; Aboulfadl, H.; Mücklich, F.; Schumacher, P. Precipitation hardening of an Mg–5Zn–2Gd–0.4Zr (wt.%) alloy. *Acta Materialia* **2016**, *108*, 207–218. [CrossRef]

20. Xia, X.; Sun, W.; Luo, A.A.; Stone, D.S. Precipitation evolution and hardening in Mg-Sm-Zn-Zr alloys. *Acta Materialia* **2016**, *111*, 335–347. [CrossRef]

21. Issa, A.; Saal, J.E.; Wolverton, C. Formation of high-strength β′ precipitates in Mg–RE alloys: The role of the Mg/β″ interfacial instability. *Acta Materialia* **2015**, *83*, 75–83. [CrossRef]

22. Liu, H.; Xue, F.; Bai, J.; Zhou, J.; Sun, Y. Effects of heat treatments on microstructures and precipitation behaviour of $Mg_{94}Y_4Zn_2$ extruded alloy. *J. Mater. Sci. Technol.* **2014**, *30*, 128–133. [CrossRef]

23. Suresh, M.; Srinivasan, A.; Pillai, U.T.S.; Pai, B.C. The effect of charcoal addition on the grain refinement and ageing response of magnesium alloy AZ91. *Mater. Sci. Eng. A* **2011**, *528*, 8573–8578. [CrossRef]

24. Ansari, M.A.; Behnagh, R.A.; Lin, D.; Kazeminia, S. Modelling of friction stir extrusion using artificial neural network (ANN). *Int. J. Adv. Des. Manuf. Technol.* **2018**, *11*, 1–12.

25. Okuyucu, H.; Kurt, A.; Arcaklioglu, E. Artificial neural network application to the friction stir welding of aluminum plates. *Mater. Des.* **2007**, *28*, 78–84. [CrossRef]

26. Asadi, P.; Givi, M.K.B.; Rastgoo, A.; Akbari, M.; Zakeri, V.; Rasouli, S. Predicting the grain size and hardness of AZ91/SiC nanocomposite by artificial neural networks. *Int. J. Adv. Manuf. Technol.* **2012**, *63*, 1095–1107. [CrossRef]

27. Yousif, Y.K.; Daws, K.M.; Kazem, B.I. Prediction of friction stir welding characteristic using neural network. *Jordan J. Mech. Ind. Eng.* **2008**, *2*, 151–155.

28. Ghetiya, N.D.; Patel, K.M. Prediction of tensile strength in friction stir welded aluminium alloy using artificial neural network. *Procedia Technol.* **2014**, *14*, 274–281. [CrossRef]

29. Arunchai, T.; Sonthipermpoon, K.; Apichayakul, P.; Tamee, K. Resistance spot welding optimization based on artificial neural network. *Int. J. Manuf. Eng.* **2014**, *2014*, 1–6. [CrossRef]

30. Habashy, D.M.; Mohamed, H.S.; El-Zaidia, E.F.M. A simulated neural system (ANNs) for micro-hardness of nano-crystalline titanium dioxide. *Physica B* **2019**, *556*, 183–189. [CrossRef]

31. Voort, G.F.; James, H.M. *Metal Handbook: Metallography and Microstructures*; American Society for Metals: Russel Township, OH, USA, 1989; Volume 9, pp. 425–434.

32. Abd El-Rehim, A.F.; Mahmoud, M.A. Transient and steady state creep of age-hardenable Al–5 wt% Mg alloy during superimposed torsional oscillations. *J. Mater. Sci.* **2013**, *48*, 2659–2669. [CrossRef]

33. Celotto, S. TEM study of continuous precipitation in Mg–9 wt% Al–1 wt% Zn alloy. *Acta Materialia* **2000**, *48*, 1775–1787. [CrossRef]

34. Abd El-Rehim, A.F. Effect of cyclic stress reduction on the creep characteristics of AZ91 magnesium alloy. *Acta Metallurgica Sinica (Engl. Lett.)* **2015**, *28*, 1065–1073. [CrossRef]

35. Zheng, M.Y.; Wu, K.; Kamado, S.; Kojima, Y. Aging behavior of squeeze cast SiCw/AZ91 magnesium matrix composite. *Mater. Sci. Eng. A* **2003**, *348*, 67–75. [CrossRef]

36. Duly, D.; Brechet, Y.; Chenal, B. Macroscopic kinetics of discontinuous precipitation in a Mg–8.5wt.% Al alloy. *Acta Metallurgica Materialia* **1992**, *40*, 2289–2300. [CrossRef]

37. Manna, I.; Pabi, S.K.; Gust, W. Discontinuous reactions in solids. *Int. Mater. Rev.* **2013**, *46*, 53–91. [CrossRef]

38. Robson, J.D. Modeling competitive continuous and discontinuous precipitation. *Acta Materialia* **2013**, *61*, 7781–7790. [CrossRef]

39. Busk, R.S. Effect of temperature on the lattice parameters of magnesium alloys. *JOM* **1952**, *4*, 207–209. [CrossRef]

40. Fatmi, M.; Djemli, A.; Ouali, A.; Chihi, T.; Ghebouli, M.A.; Belhouchet, H. Heat treatment and kinetics of precipitation of β-$Mg_{17}Al_{12}$ phase in AZ91 alloy. *Res. Phys.* **2018**, *10*, 693–698. [CrossRef]

41. Han, G.M.; Han, Z.Q.; Luo, A.A.; Liu, B.C. Microstructure characteristics and effect of aging process on the mechanical properties of squeeze-cast AZ91 alloy. *J. Alloys Compd.* **2015**, *641*, 56–63. [CrossRef]

Article

Mathematical Modeling and Analysis of Tribological Properties of AA6063 Aluminum Alloy Reinforced with Fly Ash by Using Response Surface Methodology

Alaa Mohammed Razzaq [1,2], Dayang Laila Majid [3,*], Mohamad Ridzwan Ishak [3] and Uday Muwafaq Basheer [4,5]

1   Oil Products Distribution Company, Iraq Ministry of Oil, Baghdad 10022, Iraq; alaaupm@gmail.com
2   Department of Materials Technology Engineering, Technical Engineering College-Baghdad, Middle Technical University, Baghdad 10074, Iraq
3   Department of Aerospace Engineering, Faculty of Engineering, Universiti Putra Malaysia, Serdang 43400, Malaysia; mohdridzwan@upm.edu.my
4   Faculty of Mechanical Engineering, Universiti Teknologi Malaysia, Johor Bahru 81310, Malaysia; ummb2008@gmail.com
5   Centre for Advanced Composite Materials (CACM), Institute for Vehicle Systems and Engineering, Universiti Teknologi Malaysia, Skudai 81310, Malaysia
*   Correspondence: dlaila@upm.edu.my

Received: 12 February 2020; Accepted: 1 April 2020; Published: 16 May 2020

Abstract: Lightweight, high-strength metal matrix composites have attracted considerable interest because of their attractive physical, mechanical and tribological properties. Moreover, they may offer distinct advantages due to good strength and wear resistance. In this research, AA6063 was reinforced with FA particles using compocasting methods. The effects of fly ash content, load, sliding speed and performance tribology of AA6063 –FA composite were evaluated. Dry sliding wear tests were carried out according to experimental design using the pin-on-disc method with three different loads (24.5, 49 and 73.5 N) and three speeds (150, 200 and 250 rpm) at room temperature. Response surface methodology (RSM) was used to analyze the influence of the process parameters on the tribological behavior of the composites. The surface plot showed that the wear rate increased with increasing load, time and sliding velocity. In contrast, the friction coefficient decreased with increasing these parameters. Optimal models for wear rate and friction coefficient showed appropriate results that can be estimated, hence reducing wear testing time and cost.

Keywords: AA6063; fly ash; response surface methodology; tribological properties; wear rate; friction coefficient

1. Introduction

Light-weight metal matrix composites (MMCs) have paved the way to replace steels and reduce greenhouse effects, hence they are used in various applications. A particular example is discontinuous particulate reinforced aluminum composites (DPAMCs), employed in automotive, aerospace and defense sectors [1]. DPAMCs using high-strength refractory particle reinforcements such as SiC, TiC, B_4C, Al_2O_3, MgO, TiO_2 were found to improve wear properties; attempts were undertaken to model wear behavior effectively. The modeling involves the process of evaluating the relationship between two or more process variables, as well as with the response variable. To be able to do so effectively, requires a mathematical model that is able to predict the response outcome based on various effects of the process variables [2]—especially in the evaluation of material properties. Techniques such as multiple regressions and artificial neural networks were among those applied to build mathematical model. Mechanical and tribological properties can be also predicted by employing other statistical tools, e.g., design of experiments, analysis of variance (ANOVA) and regression analysis [3–5]. Typically for wear tests, the process variables are the load, sliding speed, time and the reinforcement content.

Kumar et al. [6] investigated the size range of fly ash FA particles on dry-sliding wear behavior of aluminum composites. They observed that the wear and friction coefficient of the composites were influenced by sliding speed, reinforcement content and applied load. They developed a mathematical model for prediction of wear rate and the friction coefficient of aluminum composites reinforced with fly ash while the adequacy of the model is evaluated using ANOVA. Rajmohan et al. [7] observed that wear rate increased with increased load. They employed analysis variance ANOVA to analyze the wear behavior of the composite. They found that load is a significant factor in the wear performance of the composite. Magibalan et al. [8] reported that AA8011 composite with 12% fly ash produced lower wear rates and increased hardness with a value of 63.37 VHN. They successfully developed a regression model that can predict the wear rate of cast AA8011-12% FA composites up to 95% confidence levels, within the limits of investigation. Normal load, time and sliding velocity were shown to have a strong influence on the wear rate of the composites.

Similarly, Mandaletal et al. [9] predicted the wear behavior of metal matrix composites using regression methods and significance of wear factors using ANOVA method. The increase of parameter was also done by style trained software package. Their work suggested that response surface methodology may be suitable to evaluate the wear performance of combined rolling action and sliding. Moreover, Monikandan et al. [10] presented the tribological properties of mono AA6061-10 wt% B_4C and hybrid AA6061-10 wt% B4C-7.5 wt% Gr composites which were proposed as potential substitute for aluminum alloys used in automotive engines. A full factorial experimental design was conducted, and their findings suggested that the wear loss increased with sliding distance and applied load, while the coefficient of friction also increased with increase in applied load. In addition, the ANOVA analysis revealed the statistically and physically significant factors that influence the wear loss and friction coefficient. Processing-wise, AA6061-15%TiC aluminum matrix composites were produced by the stir casting method [11]. An empirical relationship incorporating stir casting parameters was developed for predicting ultimate tensile stress. A lower or higher combination of these parameters resulted in lower ultimate tensile stress. In this work, a compocasting technique that utilized a lower processing temperature compared to stir casting method was employed, where the lower temperature aided in reducing interfacial reactions.

The aim of the current work is to develop models to predict wear rate and coefficient of friction as functions of important wear variables. The influences of control factors such as fly ash content, load and sliding speed on wear rate and friction coefficient of AA6063-FA composites prepared by compocasting technique were experimentally evaluated considering design of experimental DOE-based central composite design (CCD) technique. Full quadratic RSM based predictive models for W_r and C_r were developed and subsequently validated with experimental results.

2. Experimental

2.1. Composites Fabrication

AA6063 alloy was selected as the matrix material. The AA6063-FA composites were fabricated by compocasting technique with wt% FA (0–12%) with the average particle size of FA particles 45 μm. The manufacturing process, metallographic of the composite, mechanical properties and its XRD spectra are available in an earlier works [12,13]. For each sample the experiment was conducted three times and the average of closely repeatable test values were registered as wear rate and coefficient of friction for composites, that it has explained in previous published papers [14,15]. Chemical composition of AA6063 and FA are presented in Tables 1 and 2 respectively [16].

Table 1. Chemical composition of tested AA6063 in wt% [16].

Elements	Cu	Fe	Mg	Mn	Si	Ti	Zn	Cr	Al
wt%	< 0.01	0.19	0.51	0.04	0.86	0.008	0.01	0.01	Balance

Table 2. Chemical composition of fly ash (FA) in wt% [16].

Elements	SiO_2	Al_2O_3	Fe_2O_3	TiO_2	K_2O	CaO	MgO	Na_2O	CuO
wt%	59.98	19.09	2.78	1.14	1.09	0.63	0.38	0.34	0.01

2.2. Statistical Analysis Using Surface Response Method SRM

The data obtained from the wear tests on the AA6063-FA composites was evaluated in Design Expert V10, Design of Experiments/ statistical analysis software using the Response Surface Method (RSM). Two models were developed, one for the wear rate and the other for the coefficient of friction. The F value is a ratio of the model and residual variance. If the ratio is near to the value of 1, then it means that the term has a remarkable influence on the response. Likewise, a *p*-value larger than 0.10 indicated that the term has an insignificant effect [17].

The analysis of variance ANOVA technique was employed in analyzing the wear performance of AA6063-FA composites subjected to important controlled parameters. In current study FA was observed as a remarkable contributor in the wear performance of the composites. The robustness of the developed model is typically checked using ANOVA technique [18–20]. In this technique, the F ratio of the developed model that was computed must be lower than the standard tabulated value of F ratio for a desired level of confidence (say 95%) in order for the model to be satisfactory within that particular confidence limit. Furthermore, the coefficient of correlation 'r' was utilized in finding the closeness of predicted and experimental values which is described by the following Equations described by Kumar and Balasubramanian [18]:

$$r^2 = \frac{\text{explained variation}}{\text{total variation}} = \frac{\Sigma\left(W_{rp} - W_{ra}\right)^2}{\Sigma(W_{re} - W_{ra})^2} \tag{1}$$

$$r^2 = \frac{\text{explained variation}}{\text{total variation}} = \frac{\Sigma\left(F_{rp} - F_{ra}\right)^2}{\Sigma(F_{re} - F_{ra})^2} \tag{2}$$

where W_{rp}, F_{rp} are the predicted wear rate and coefficient of friction respectively for the given experimental wear rate, W_{re} and coefficient of friction F_{re}, for the corresponding average of experimental wear rate W_{ra} *and* average coefficient of friction F_{ra} respectively.

2.3. Experimental Design

RSM was used to model the wear rate and friction coefficient of AA6063-FA composite material. The response function representing the variables, namely, FA content, sliding speed and applied load can be expressed as [21,22] Y = f (FA, S, L). The response surface Y for the k parameter is represented in a second order polynomial regression equation, presented as the following:

$$Y = b_0 + \sum_{i=1}^{k} b_i x_i + \sum_{\substack{i=1 \\ i \neq j}}^{k} b_i x_i^2 + \sum_{i \neq j}^{k} b_{ij} x_i x_j \tag{3}$$

Expanding the polynomial for three factors will give the equation:

$$Y = b_0 + b_1 FA + b_2 L + b_3 S + b_{11} FA^2 + b_{22} L^2 + b_{33} S^2 + b_{12} FAL + b_{13} FAS + b_{23} LS \tag{4}$$

where b_0 is the average of responses and b_1, b_2, b_3, b_{12}, b_{13} *and* b_{23} are the coefficients of response dependent upon the respective main and interaction effects of parameters.

From the factorial design, the combined effects of all parameters will result in an empirical relationship between the wear rate and coefficient of friction with the designated parameters which are the fly ash content (*FA*), load (*L*) and sliding speed (*S*). The multiple regression technique used to develop the mathematical model to formulate the influence of the parameters on both responses were achieved via design of experiments analysis software, Design-Expert V10.0.

Feasible limits of all three factors were determined and these are presented in Table 3.

Table 3. Levels of factors.

No.	Factor	Notation	Unit	Levels						
1	Fly ash content	FA	wt%	0	2	4	6	8	10	12
2	Load	L	N	24.5	49	73.5	–	–	–	–
3	Sliding speed	S	Rpm	150	200	250	–	–	–	–

There were 63 experimental response values of various combination of factors obtained and used to formulate the response function (wear rate and coefficient of friction) as shown in Table A1 (Appendix A) and also as input into the Design-Expert V10.0 software. The highlighted results (Appendix A), which represent 20% of the total data, was not used in the model construction process. This data were utilized in verification of the proposed empirical model. The rest of the data were introduced in full factorial design using response surface method with user-defined choice. The software then automatically will evaluate all possible combinations of the factors and produce results for the significant and aliased model terms.

The design summary of the problem is tabulated in Table 4, which shows the summary of the input information and setting required to starting the analysis. In summary, the methods of the current work encompass a quantitative (wear test) and statistical analyses (randomized RSM approach), whereby the statistical method provide a means to assess the effect of all three factors, namely the fly ash content (*FA*), load (*L*) and sliding speed (*S*) on the wear properties with a quadratic design model.

Table 4. Factorial design summary (Design-Expert V10.0).

Factor	Name	Units	Type	Subtype	Min	Max	Coded Values		Mean	Std. Dev.
A	Fly ash content	wt%	Numeric	Discrete	0	12	−1.000 = 0	1.000 = 12	6.15686	4.05646
B	Sliding speed	Rpm	Numeric	Discrete	150	250	−1.000 = 150	1.000 = 250	197.059	40.5114
C	Load	N	Numeric	Discrete	24.5	73.5	−1.000 = 24.5	1.000 = 73.5	49.4804	19.898

Response	Name	Units	Obs	Analysis	Min	Max	Mean	Std. Dev.	Ratio	Trans	Model
R1	Wear rate	mm^3/m	51	Polynomial	3.4623	13.3471	7.63128	2.44359	3.85498	None	Quadratic
R2	Coefficient of friction		51	Polynomial	0.1085	0.4723	0.236767	0.0830769	4.353	None	Quadratic

3. Results and Discussion

3.1. Statistical Analyses

In this section, the effect of FA content, sliding speed and applied load on wear rate and coefficient of friction are discussed. RSM was applied to evaluate and develop models of wear rate and friction coefficient of AA6063-FA composites.

In an attempt to obtain a model with better performance than the one given by the quadratic model of the process, a cubic model equation was developed, but owing to the fact that the developed cubic model was found to be aliased in nature, its model expression in terms of the actual factors was not given by the Design Expert. However, the aliased cubic model, the mathematical expression of which was not available, was still analyzed for variance to see if actually increasing order of model had any effect on the response statistically.

The models summary statistics given in Tables 5 and 6 show that the quadratic models were suggested according to the statistical evaluation of different models. The software also revealed that the cubic order model was aliased for the ranges of the data obtained. The quadratic model was used according to the value of R-squared (0.9791, 0.9818) for wear rate and coefficient of friction model respectively.

Table 5. Statistical summary of wear rate model.

Source	Std. Dev.	R-Squared	Adjusted R-Squared	Predicted R-Squared	PRESS	Recommendation
Linear	0.60	0.9437	0.9401	0.9338	19.76	
2FI	0.48	0.9656	0.9609	0.9548	13.50	
Quadratic	0.39	0.9791	0.9745	0.9658	10.20	Suggested
Cubic	0.36	0.9858	0.9784	0.9633	10.97	Aliased

Table 6. Statistical Summary of Coefficient of Friction Model.

Source	Std. Dev.	R-Squared	Adjusted R-Squared	Predicted R-Squared	PRESS	Recommendation
Linear	0.024	0.9214	0.9164	0.9043	0.033	
2FI	0.020	0.9483	0.9412	0.9242	0.026	
Quadratic	0.012	0.9818	0.9778	0.9699	0.010	Suggested
Cubic	0.009406	0.9915	0.9872	0.9769	0.007958	Aliased

3.2. Modeling and Analysis of Variance of Wear Rate

The design of experiment software with response surface method was used to statistically study the effect of fly ash (FA), sliding speed (S) and applied load (L) on the wear rate of AA6063-FA composites and to build an empirical model for Wear rate based on the effects of these factors. The sequential F-test was conducted for testing the significance of the regression model.

The step-wise regression method was utilized for the wear rate model with the entire potential combinations of the control factors with exception of the cubic terms since these terms were aliased for the data ranges obtained. The selected terms along with the step-wise regression method led to the elimination of the less significant model terms automatically. Table 7 shows the analysis of variance generated for wear rate model. The model F-value of 213.395 confirms the significance of the model. A p-value of 0.0001 was indicative that there was a low chance of F-value which could occur as result of noise. The predicted R^2 of 0.9658, which represent the measure of the amount of variation about the mean explained by the model agrees well with the adjusted R^2 of 0.9745. The adjusted R^2 and predicted R^2 should be within approximately 0.20 of each other so as to be in agreement, or else setbacks may arise with either the data or the model. High R^2 value suggest there is a statistically significant interaction between factors. Resulting wear rate within the investigated ranges of parameters can be modeled by the final equation produced, Equation (5).

$$\begin{aligned} W_r = {} & 0.037685 - 0.10651 \times FA + 0.054471 \times S + 0.019936 \times L - 0.001056 \times FA \times S \\ & - 0.00337143 \times FA \times L + 0.0000796096 \times S \times L + 0.0070058 \times FA^2 - 0.0000709692 \times S^2 + \\ & 0.000880987 \times L^2 \end{aligned} \quad (5)$$

The order of the factors influencing the wear rate can be established through analysis of the F-value magnitude as follows: it can be observed the F-value of FA (837.432) was the most significant parameter having the highest statistical influence on the wear rate of AA6063-FA composites followed by applied load (483.037) and sliding speed (286.805) and S, which implies that the FA content has greater effect on wear rate value than the L and S factor.

Adequate precision ratio, which represent the signal-to-noise ratio, with values higher than 4 indicate adequate model discrimination [17]. It compares the range between the predicted values at the design points to the average prediction error. As shown in Table 7, the adequate ratio was 59.006, which indicates that the model has capacity of navigating the design space. The R^2 value of about 0.9791 indicates the variability of the response was 97.91% about the mean which proved the model provide a good fit for the data. This fact alongside with the remarkable residual analysis led to the conclusion that the model given in Equation (5) can predict the wear rate of the compound within the investigated range of all three parameters.

Table 7. ANOVA results for the quadratic model of wear rate.

Source	Sum of Squares	df	Mean Square	F Value	*p*-Value Prob > F	
Model	292.32	9	32.48	213.39	<0.0001	Significant
A-weight fraction	127.46	1	127.46	837.43	<0.0001	
B-sliding speed	43.65	1	43.65	286.80	<0.0001	
C-Load	73.52	1	73.52	483.04	<0.0001	
AB	1.36	1	1.36	8.96	0.0047	
AC	3.49	1	3.49	22.95	<0.0001	
BC	0.20	1	0.20	1.30	0.2600	
A^2	0.48	1	0.48	3.18	0.0819	
B^2	0.35	1	0.35	2.33	0.1348	
C^2	3.17	1	3.17	20.83	<0.0001	
Residual	6.24	41	0.15			
Cor Total	298.56	50				

Std. Dev.		0.39	R-squared		0.9791
Mean		7.63	Adj R-squared		0.9745
C.V. %		5.11	Pred R-squared		0.9658
PRESS		10.20	Adequate Precision		59.006

Furthermore, in Table 7, both the *SL* (BC) and S^2 (B^2) terms are considered to be insignificant as their *p*-value was greater than 0.05. However, these terms were automatically added in order to support the model's hierarchy. In contrast, it is evident that the combined interaction of FA×L has the strongest effect on the response. The essential effects of the parameters on the responses, i.e., wear rate and the coefficient of friction, are further illustrated using contour lines and 3D surface plots.

Graphical Results of Wear Rate Model

Figure 1 presents the variance of collected data points about the linearized line which relates the predicted values with the actual values. This observation clearly indicates the model provide a good fit for the wear rate response.

Figure 2 further presents the perturbation plot of the wear rate data at the experimented mid values of the process parameters. The comparison of the effects of all three parameters at a specific area in the design range can be enabled by the perturbation plot. The response (wear rate) was graphically plotted by changing a single parameter over its investigated limits while other parameters were maintained constant. Similar to the significance of the *p*-value, this tool was also effective in identifying the interactive effects of parameters on the response [17]. The X-axis of the plot showed the relative position of the chosen levels of the parameters to the coded scale. The slopes of the curves showed the rate at which the parameter influenced the response. In Figure 2, the point selected in the design range was the central point (FA = 6 wt%, S 200 rpm and L 49 N).

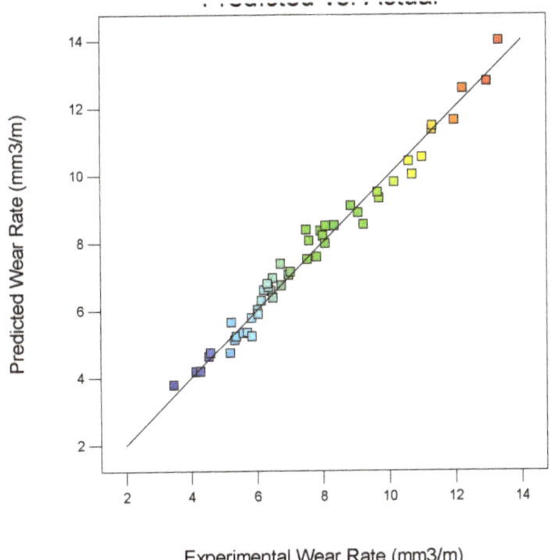

Figure 1. Experimental versus predicted wear rate mm^3/m.

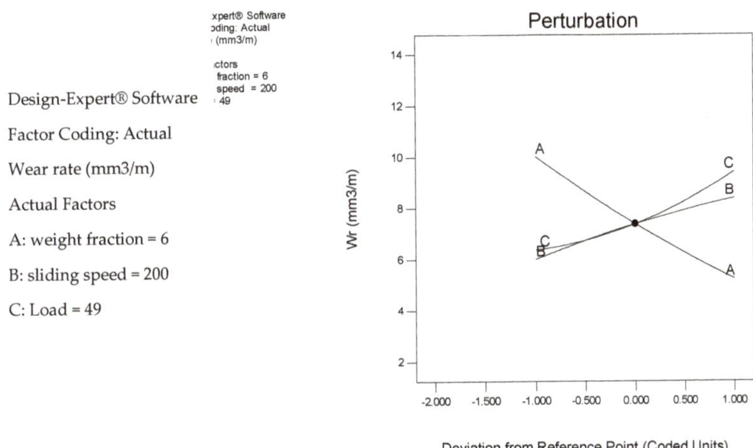

Design-Expert® Software

Factor Coding: Actual

Wear rate (mm3/m)

Actual Factors

A: weight fraction = 6

B: sliding speed = 200

C: Load = 49

Figure 2. Perturbation plot of the effect of Fly ash content, sliding speed and applied load on wear rate of AA6063-FA composites.

The plot indicates that the fly ash content FA has the strongest inverse proportion (negative slope) on the wear rate while the load L has a smaller effect than FA, which was a direct proportional one (positive slope). The quadratic curves depicting the influence of the parameters on the wear rate were attributed to the quadratic terms in the model.

In summary, it is evident from Figure 2 that the FA content has negative effect on wear rate while both the applied load and sliding speed exhibit a positive impact on the wear rate. By increasing the FA content, the wear rate can be decreased while the opposite can be obtained when the applied load and sliding speed were increased. This trend was similar to that obtained from the experimental results.

The combined effects of two process parameters on the wear rate, while the third was kept constant, were further evaluated and these are demonstrated in Figures 3–5. As described earlier for the perturbation plot, the third parameter could be changed, but again, it was kept at similar value selected for the perturbation figure. The red spheres on the graphs represent the experimental data points. Moreover, it is evident, especially on the 3D plots, that these spheres have close location to the surfaces, which indicate that the model fits properly with the data points. With respect to the effects of the combined parameters, they portrayed similar findings to those presented in the perturbation plot. However, by identifying their effects relative to the numerical values of the response, they offer a visual aid in selecting the desired ranges of the parameters. Furthermore, they were important in the optimization process of the ranges of the processing parameters so as to meet certain response criteria. Owing to these facts and to avoid redundancy, only brief discussions were given for each figure as provided in the following paragraph.

Figure 3a shows the contour plot of the effect of S and FA on the wear rate Wr. Each contour curve showed combinations of the two parameters that will predict a constant wear rate value. This wear rate value was observed inside the box situated at the middle of each contour curve. From the contour plot, S has a direct proportional effect on the predicted wear rate, while FA was inversely proportionate. The same data are further illustrated in 3D plot as shown in Figure 3b. The 3D plot has the added advantage of visual display of the effect of one of the parameter changes when the value of the other parameter is modified. For instance, consider the effect of S at two different values of FA which were at 0- and 12-wt%, it can be observed that the FA effect was stronger in the first case.

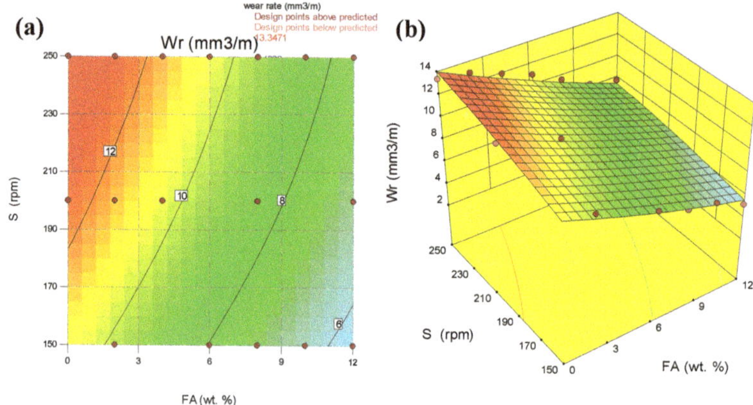

Figure 3. View of the interactive effect of FA and S on Wr, (**a**) contour lines 2D, (**b**) 3D.

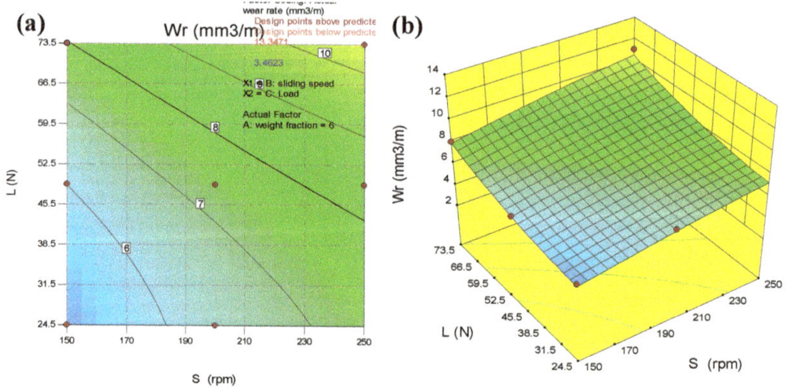

Figure 4. View of the interactive effect of S and L on the Wr (**a**) contour lines 2D, (**b**) 3D.

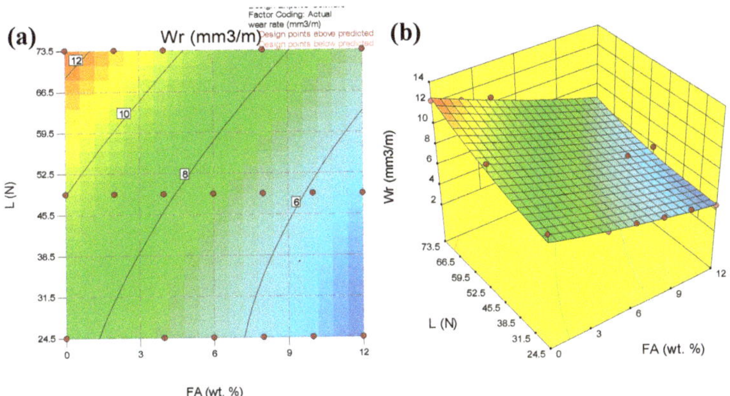

Figure 5. FA and L contour showing the interactive effect on Wr, (**a**) contour lines, (**b**) 3D.

Figure 4a,b present that both S and L have directly proportional effects on the Wr. However, L has stronger effect than S. This indicated that variations in L led to larger variations in the Wr than those emanating from large variations in S.

Figure 5a,b shows that FA has an inversely proportional effect on the Wr, while L has a directly proportional effect. Similarly, FA has stronger effect on the response than L exhibits.

3.3. Modeling and Analysis of Variance for Coefficient of Friction

Likewise, the previous analyses described in Section 3.2 were repeated for the case of the coefficient of friction response. Again, the step-wise regression method was employed with the entire potential combinations of all three parameters and similarly too, the cubic terms were aliased for the ranges of the data obtained. The selection of the terms alongside the step wise regression method resulted in the elimination of the negligible model terms automatically [17]. Table 8 summarizes the computed ANOVA results with the variance for the Coefficient of friction model and all terms in the model.

Table 8. ANOVA for Response Surface Quadratic model (coefficient of friction model).

Source	Sum of Squares	df	Mean Square	F Value	*p*-Value Prob > F	
Model	0.34	9	0.038	245.70	< 0.0001	Significant
A-Fly ash content	0.19	1	0.19	1219.47	< 0.0001	
B-Load	0.069	1	0.069	451.39	< 0.0001	
C-Sliding speed	0.061	1	0.061	399.07	< 0.0001	
AB	6.376×10^{-3}	1	6.376×10^{-3}	41.61	< 0.0001	
AC	1.054×10^{-3}	1	1.054×10^{-3}	6.88	0.0122	
BC	5.305×10^{-4}	1	5.305×10^{-4}	3.46	0.0700	
A^2	0.010	1	0.010	66.91	< 0.0001	
B^2	1.049×10^{-3}	1	1.049×10^{-3}	6.85	0.0124	
C^2	2.273×10^{-5}	1	2.273×10^{-5}	0.15	0.7021	
Residual	6.282×10^{-3}	41	1.532×10^{-4}			
Cor Total	0.35	50				
Std. Dev.	0.012			R-squared		0.9818
Mean	0.24			Adj R-squared		0.9778
C.V.%	5.23			Pred R-squared		0.9699
PRESS	0.010			Adeq Precision		66.605

In Table 8, it is shown that combined interaction of FA × L also had the strongest effects on the coefficient of friction. The effect of the parameters on the response is further demonstrated via the contour and 3D surface plots. The S² and SL terms were proven to be insignificant as evidenced from the *p*-value which was greater than 0.05. Nonetheless, this term again was also to support the model hierarchy. The step-wise regression method detected that this model term was insignificant however, it was necessary by other significant model terms. Moreover, in Table 8, the adequacy measures of R^2, adjusted R^2 and predicted R^2 were near to 1; they all agree well which is indicative of sufficient model [23]. The adequacy precision was greater than 4 which indicated sufficient model discrimination. In addition, this shows that the model has the capacity of navigating the design space [17]. The R^2 value of approximately 0.9818 implies that around 98.18 of the data variability can be substantiated by model. Again, similarly to the wear rate, the model was proven to be a very good fit to the data and that the coefficient of friction, within the range of parameters investigated, can be predicted. Equation (6) presents the process model estimated from the experimental results within the range of parameters investigated.

$$F_r = 0.74823 - 0.040431 \times FA - 0.00516264 \times L - 0.00103544 \times S + 0.000144033 \times FA \times L +$$
$$0.0000293612 \times FA \times S + 0.00000411498 \times S \times L +$$
$$0.00101923 \times FA^2 + 0.0000160301 \times L^2 + 0.00000056855 \times S^2$$
$$(6)$$

In summary, the model described by Equation (6) can provide good predictions of the coefficient of friction through any combinations of the fly ash content, sliding speed and load, within the range investigated.

Graphical Results of Coefficient of Friction

Figure 6 presents the variance of data points about the linearized line relating the actual and predicted values, which is an indication of the good fitness value of the model.

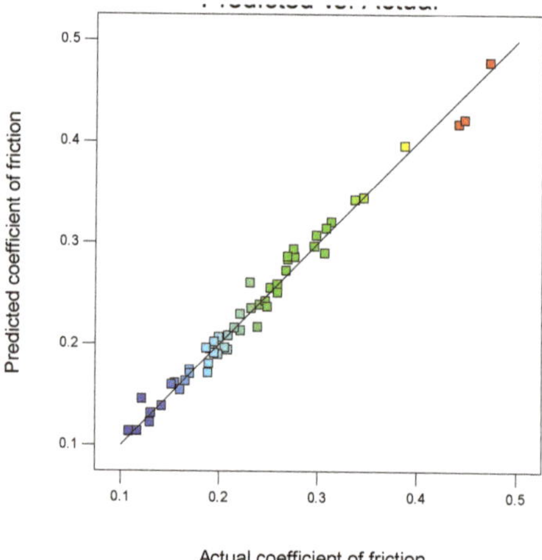

Figure 6. Actual versus predicted scattering of the data points.

For the process parameters, the perturbation (see Figure 7) was obtained at a similar point of location (FA = 6 wt%, S 200 rpm and L 49 N). From the figure, it is evident that FA has inversely proportional effect and has the maximum effect on the coefficient of friction. The second maximum effect was owing to L which also had an inversely proportional effect on the coefficient of friction. The sliding speed also exhibited an inversely proportional effect.

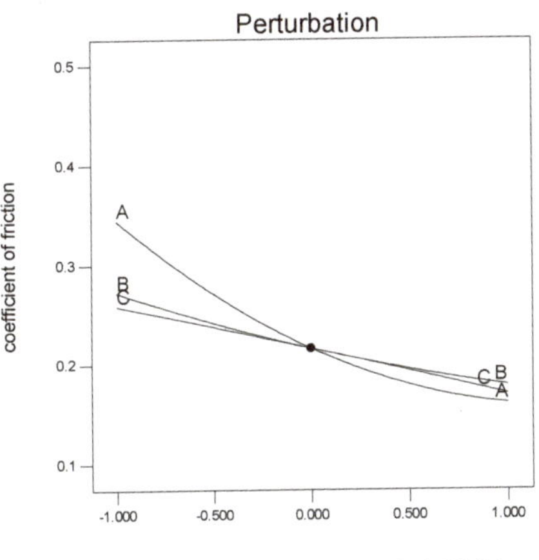

Figure 7. Perturbation plot of the process control variables' effects on the coefficient of friction.

The following Figures 8–10, present the combined influences of two parameters, at a time on F_r. the design points are represented by the red spheres on the graphs; their locations on the 3D plots was a confirmation that both the model and the data points have a good fit. The contour plot for the influence of FA and S on F_r is presented in Figure 8a. It is evident from the contour plot that S and FA have an inversely proportional effect on F_r. It can be observed from the 3D plot in Figure 8b that as FA and S decrease the F_r decreases. By employing the combination of low S and low FA settings, the deepest F_r values were generated, while holding L constant, in this case at its mid-range = 49 N.

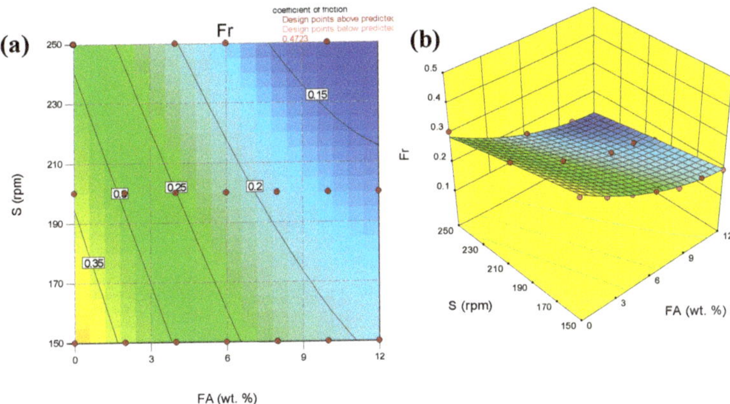

Figure 8. View of the interactive effect of FA and S on Fr, (**a**) contour lines 2D, (**b**) 3D.

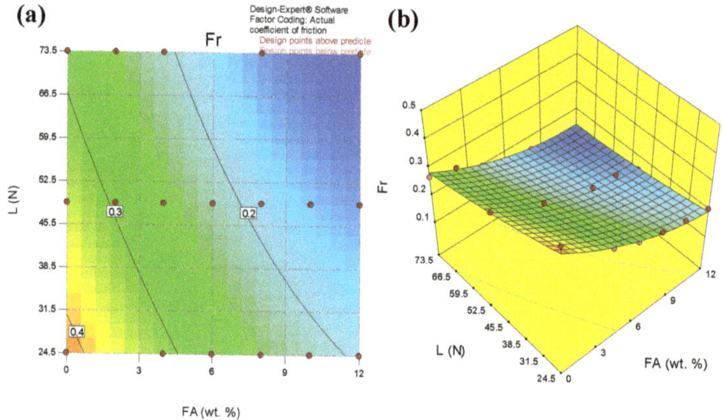

Figure 9. FA and L contour showing the interactive effect on the Fr, (**a**) contour lines 2D, (**b**) 3D.

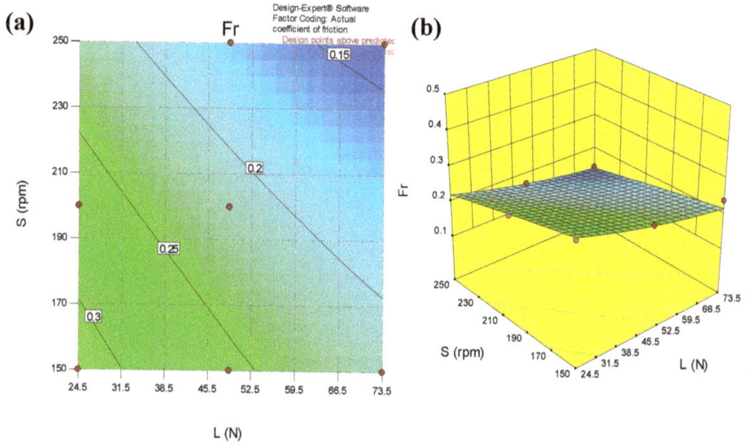

Figure 10. View of the interactive effect of S and L on F_r, (**a**) contour lines 2D, (**b**) 3D.

Figure 9a,b indicates that both FA and L have an inversely proportional influence on the F_r. Nonetheless, FA exhibited stronger effect than L. Slight changes in FA often affect the F_r in a magnitude greater than changes caused by even large changes in L.

Figure 10a,b shows that S and L have an inversely proportional effect on the F_r. In a similar way, S, which can be evidenced from the slopes of the contour lines and 3D surface, almost exhibited equal to L effect on the F_r.

3.4. Validation of the Statistical Models

The wear rate and coefficient of friction were numerically modeled by RSM based on FA content, applied load and sliding speed. Thereafter, the statistical model was validated with a dataset, 20% from the experimental data were used in the process of models building to verify the produced models and find the reliability of the model and the measured responses were recorded. The predicted wear rate and coefficient of friction calculated by Equations (5) and (6) were plotted against experimental data in Figures 11 and 12. The figures indicated that the surface response method model provides excellent coincidence with experimental value with R-squared value of 0.9665 and

0.9893 respectively. These results were expected due to the high order model used in SRM which provide more accurate response.

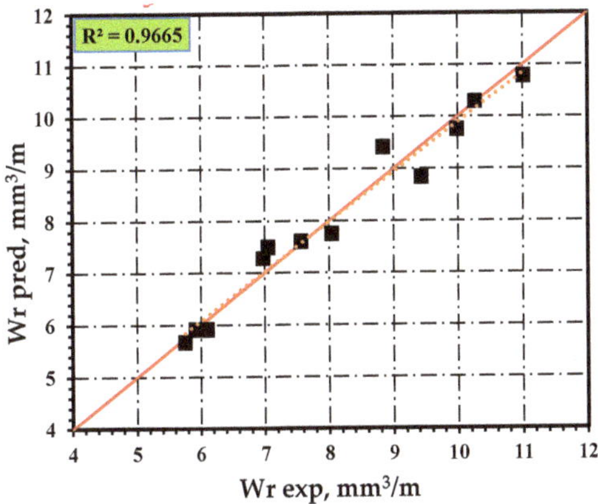

Figure 11. SRM-predicted wear rate versus experimental results.

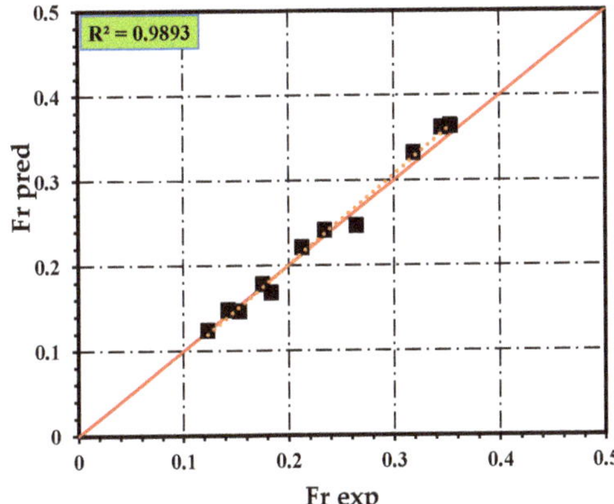

Figure 12. SRM-predicted coefficient of friction versus experimental results.

4. Conclusions

AA6063 alloy-based aluminum matrix composites with FA particles have been successfully fabricated with compocasting technique. The overall results revealed that the wear resistance of the composites was positively enhanced by the inclusion of the FA particles. The wear resistance enhancement was attributed to increased hardness, generation of strain fields, homogenous distribution, spherical shape of FA particles and reduction in effective contact area. Increasing FA content restricted the deformation of the matrix material. The wear rates of the composites were increased by the applied load. FA content exhibited an inversely proportional relationship with wear rate, coefficient of friction and friction force for AA6063-FA composites. Increasing the sliding speed increased the wear rate but decreased the coefficient of friction and friction force of composites. In addition, the wear rate and fractional force of composites increased with increasing applied load, but coefficient of friction varied inversely with applied load. These control parameters were then employed to build statistical models for the wear rate and coefficient of friction. FA had the highest statistical influence on the wear rate and coefficient of friction of AA6063-FA composites followed by applied load and sliding speed. Moreover, FA content with applied load exhibited the strongest interaction effects. Finally, AA6063 aluminum matrix composites reinforced with FA can be used in applications that require excellent wear resistance.

Author Contributions: The research conceptualization and methodology are contributed by A.M.R., D.L.M. and U.M.B.; the experimental and numerical works including the data curation are performed by A.M.R. and U.M.B.; the results analysis and interpretation are conducted by A.M.R., D.L.M., U.M.B. and M.R.I.; the preparation of the original draft, reviews and editing are contributed by A.M.R. and D.L.M. and lastly the project administration, supervision and funding are managed by D.L.M. All authors have read and agreed to the published version of the manuscript.

Funding: The authors would like to acknowledge the financial support from the Ministry of Education Malaysia with reference code FRGS/1/2019/STG07/UPM/02/10 and Vot No. 5540209.

Conflicts of Interest: The authors declare no conflict of interest. The funders had no role in the design of the study; in the collection, analyses, or interpretation of data; in the writing of the manuscript, or in the decision to publish the results.

Nomenclature

min	Minute
rpm	Revolution per minute
ASTM	American Society for Testing and Materials
FA	Fly ash
Al	Aluminum
AA6063	AA6063 aluminum alloy
AA6063-FA	AA6063 aluminum alloy-fly ash composites
MMCs	Metal matrix composites
AMCs	Aluminum matrix composites
AMMCs	Aluminum metal matrix composites
Al-FA	Aluminum–fly ash
B4C	Boron carbide
Graphite	Gr
XRD	X-ray diffraction
XRF	X-ray fluorescence
ANOVA	Analysis of variance
RSM	Response surface method
DRAMCs	Discontinuously reinforced aluminum matrix composites
DOE	Design of experimental
CCD	Based central composite design
SS	Semisolid-semisolid
SL	Semisolid-liquid
Wr	Wear rate (mm3/m)
Fr	Coefficient of friction
μ	The coefficient of friction
FT	Tangential force (N)
FN	Normal force (N)
V	Volume loss (mm3)
S	Sliding speed (rpm)
L	Applied load (N)
Wrp	Predicted wear rate
Frp	Predicted coefficient of friction
Wre	Experimental wear rate
Fre	Experimental coefficient of friction
Wra	Average of experimental wear rate
Fra	Average of experimental coefficient of friction
vol%	Percentage of volume fraction
wt%	Percentage of weight fraction

Appendix A

Table A1. Parameters used in mathematical models for wear analysis.

NO	FA wt%	Load N	Sliding Speed Rpm	Wear Rate mm^3/m	Coefficient of Friction
1	0	24.5	150	6.4867	0.4723
2	0	24.5	200	9.25	0.4471
3	0	24.5	250	9.9889	0.3451
4	0	49	150	7.9484	0.3868
5	0	49	200	10.7283	0.346
6	0	49	250	11.3323	0.3069
7	0	73.5	150	11.0173	0.3187
8	0	73.5	200	12.2625	0.269
9	0	73.5	250	13.3471	0.2486
10	2	24.5	150	6.136	0.4414
11	2	24.5	200	8.0448	0.3541
12	2	24.5	250	9.0934	0.2985
13	2	49	150	7.5484	0.3372
14	2	49	200	8.875628	0.2755
15	2	49	250	10.2673	0.2344
16	2	73.5	150	10.1834	0.2766
17	2	73.5	200	11.3503	0.2468
18	2	73.5	250	12.9839	0.1873
19	4	24.5	150	5.7524	0.3533
20	4	24.5	200	6.9738	0.3086
21	4	24.5	250	7.6022	0.2313
22	4	49	150	6.7511	0.2963
23	4	49	200	8.0254	0.2592
24	4	49	250	9.7357	0.1954
25	4	73.5	150	9.4389	0.2647
26	4	73.5	200	10.6226	0.2001
27	4	73.5	250	12.0139	0.166
28	6	24.5	150	5.3246	0.3135
29	6	24.5	200	6.504	0.2676
30	6	24.5	250	6.974	0.2129
31	6	49	150	6.0267	0.2589
32	6	49	200	6.7292	0.2155
33	6	49	250	7.5166	0.1707
34	6	73.5	150	8.0876	0.2391
35	6	73.5	200	8.8426	0.1754
36	6	73.5	250	11.0318	0.1415
37	8	24.5	150	4.5472	0.2694
38	8	24.5	200	5.83851	0.2409
39	8	24.5	250	6.4298	0.1954
40	8	49	150	5.7058	0.2214
41	8	49	200	6.2207	0.1991
42	8	49	250	7.0522	0.1535
43	8	73.5	150	7.0393	0.209
44	8	73.5	200	8.1105	0.1518
45	8	73.5	250	9.6893	0.1296
46	10	24.5	150	4.1575	0.2518
47	10	24.5	200	5.3748	0.2216
48	10	24.5	250	5.9217	0.1835
49	10	49	150	5.1829	0.2093
50	10	49	200	6.0414	0.1891
51	10	49	250	6.3559	0.1306
52	10	73.5	150	6.5035	0.1899
53	10	73.5	200	7.5705	0.1427
54	10	73.5	250	8.367	0.1167
55	12	24.5	150	3.4623	0.2324

Table A1. *Cont.*

NO	FA wt%	Load N	Sliding Speed Rpm	Wear Ratemm^3/m	Coefficient of Friction
56	12	24.5	200	4.598	0.2059
57	12	24.5	250	5.58011	0.1607
58	12	49	150	4.2891	0.1912
59	12	49	200	5.8519	0.1553
60	12	49	250	6.1012	0.1234
61	12	73.5	150	5.221376	0.1701
62	12	73.5	200	6.3248	0.1217
63	12	73.5	250	7.8319	0.1085

References

1. Sardar, S.; Pradhan, S.K.; Karmakar, S.K.; Das, D. Modeling of abraded surface roughness and wear resistance of aluminum matrix composites. *J. Tribol.* **2019**, *141*, 071601. [CrossRef]

2. Nouri, M.; Abdollah-Zadeh, A.; Malek, F. Effect of welding parameters on dilution and weld bead geometry in cladding. *J. Mater. Sci. Technol.* **2007**, *23*, 817.

3. Palanivel, R.; Mathews, P.K.; Murugan, N. Development of mathematical model to predict the mechanical properties of friction stir welded AA6351 aluminum alloy. *J. Eng. Sci. Technol. Rev.* **2011**, *44*, 25–31. [CrossRef]

4. Heidarzadeh, A.; Saeid, T.; Khodaverdizadeh, H.; Mahmoudi, A.; Nazari, E. Establishing a mathematical model to predict the tensile strength of friction stir welded pure copper joints. *Metall. Mater. Trans. B* **2013**, *44*, 175–183.

5. Lakshminarayanan, A.; Balasubramanian, V. Comparison of RSM with ANN in predicting tensile strength of friction stir welded AA7039 aluminium alloy joints. *Trans. Nonferrous Met. Soc. China* **2009**, *19*, 9–18. [CrossRef]

6. Kumar, K.R.; Mohanasundaram, K.M.; Arumaikkannu, G.; Subramanian, R. Analysis of parameters influencing wear and frictional behavior of aluminum—Fly ash composites. *Tribol. Trans.* **2012**, *55*, 723–729. [CrossRef]

7. Rajmohan, T.; Palanikumar, K.; Ranganathan, S. Evaluation of mechanical and wear properties of hybrid aluminium matrix composites. *Trans. Nonferr. Met. Soc. China* **2013**, *23*, 2509–2517. [CrossRef]

8. Magibalan, S.; Senthilkumar, P.; Palanivelu, R.; Senthilkumar, C.; Shivasankaran, N.; Prabu, M. Dry sliding behavior of aluminum alloy 8011 with 12% fly ash composites. *Mater. Res. Express* **2018**, *5*, 056505. [CrossRef]

9. Mandal, N.; Roy, H.; Mondal, B.; Murmu, N.C.; Mukhopadhyay, S. Mathematical modeling of wear characteristics of 6061 Al-alloy-SiCp composite using response surface methodology. *J. Mater. Eng. Perform.* **2012**, *21*, 17–24.

10. Monikandan, V.V.; Jacob, J.C.; Joseph, M.A.; Rajendrakumar, P. Statistical analysis of tribological properties of aluminum matrix composites using full factorial design. *Trans. Indian Inst. Met.* **2015**, *68*, 53–57. [CrossRef]

11. Moses, J.J.; Dinaharan, I.; Jekhar, S.J. Prediction of influence of process parameters on tensile strength of AA6061/TiC aluminum matrix composites produced using stir casting. *Trans. Nonferr. Met. Soc. China* **2016**, *26*, 1498–1511. [CrossRef]

12. Razzaq, A.M.; Majid, D.L.; Ishak, M.R.; Uday, M.B. Microstructural characterization of fly ash particulate reinforced AA6063 aluminium alloy for aerospace applications. *Mater. Sci. Eng. Conf. Ser.* **2017**, *270*, 1.

13. Razzaq, A.M.; Majid, D.L.; Ishak, M.R.; Uday, M.B. Influence of fly ash on the microstructure and mechanical properties of AA6063 alloy using compocasting technique. *Mater. Express* **2019**, *9*, 1–14. [CrossRef]

14. Razzaq, A.M.; Majid, D.L.; Manan, N.H.; Ishak, M.R.; Uday, M.B. Effect of fly ash content and applied load on wear behaviour of AA6063 aluminium alloy. In *IOP Conference Series: Materials Science and Engineering*; IOP Publishing: Bristol, UK, 2018.

15. Razzaq, A.M.; Majid, D.L.; Ishak, M.R.; Uday, M.B. *Effects of Solid Fly Ash on Wear Behaviour of AA6063 Aluminum Alloy*; Elsevier: Amsterdam, The Netherlands, 2019.

16. Razzaq, A.M.; Majid, D.L.; Ishak, M.R.; Uday, M.B. Effect of fly ash addition on the physical and mechanical properties of AA6063 alloy reinforcement. *Metals* **2017**, *77*, 477. [CrossRef]

17. Design-Expert Software. *User's Guide, Technical Manual*; Stat-Ease Inc.: Minneapolis, MN, USA, 2000; Volume 6.

18. Kumar, S.; Balasubramanian, V. Effect of reinforcement size and volume fraction on the abrasive wear behaviour of AA7075 Al/SiC p P/M composites—A statistical analysis. *Tribol. Int.* **2010**, *43*, 414–422. [CrossRef]

19. Balasubramanian, V.; Lakshminarayanan, A.K.; Varahamoorthy, R.; Babu, S. Application of response surface methodolody to prediction of dilution in plasma transferred arc hardfacing of stainless steel on carbon steel. *J. Iron Steel Res. Int.* **2009**, *16*, 44–53. [CrossRef]

20. Elangovan, K.; Balasubramanian, V.; Babu, S. Predicting tensile strength of friction stir welded AA6061 aluminium alloy joints by a mathematical model. *Mater. Des.* **2009**, *30*, 188–193. [CrossRef]

21. Gopalakrishnan, S.; Murugan, N. Production and wear characterisation of AA 6061 matrix titanium carbide particulate reinforced composite by enhanced stir casting method. *Compos. Part B Eng.* **2012**, *43*, 302–308. [CrossRef]

22. Kumar, S.; Balasubramanian, V. Developing a mathematical model to evaluate wear rate of AA7075/SiC p powder metallurgy composites. *Wear* **2008**, *264*, 1026–1034. [CrossRef]

23. Badkar, D.S.; Pandey, K.S.; Buvanashekaran, G. Development of RSM-and ANN-based models to predict and analyze the effects of process parameters of laser-hardened commercially pure titanium on heat input and tensile strength. *Int. J. Adv. Manuf. Technol.* **2013**, *65*, 1319–1338. [CrossRef]

Article

Hot Deformation Behavior and Microstructure Characterization of an Al-Cu-Li-Mg-Ag Alloy

Lingfei Cao [1,2], Bin Liao [3], Xiaodong Wu [1,2,*], Chaoyang Li [1], Guangjie Huang [1] and Nanpu Cheng [4,*]

1   International Joint Laboratory for Light Alloys (Ministry of Education),
    College of Materials Science and Engineering, Chongqing University, Chongqing 400044, China;
    caolingfei@cqu.edu.cn (L.C.); licy@cqu.edu.cn (C.L.); gjhuang@cqu.edu.cn (G.H.)
2   Shenyang National Laboratory for Materials Science, Chongqing University, Chongqing 400044, China
3   Alnan Aluminium Co., Ltd., Nanning Guangxi 530031, China; bin.liao@foxmail.com
4   School of Materials and Energy, Southwest University, Chongqing 400715, China
*   Correspondence: xiaodongwu@cqu.edu.cn (X.W.); cheng_np@swu.edu.cn (N.C.);
    Tel.: +86-23-65127306 (X.W.); +86-23-68253204 (N.C.)

Received: 19 April 2020; Accepted: 20 May 2020; Published: 22 May 2020

Abstract: The flow behavior of an Al-Cu-Li-Mg-Ag alloy was studied by thermal simulation tests at deformation temperatures between 350 °C and 470 °C and strain rates between 0.01–10 s^{-1}. The microstructures of the deformed materials were characterized by electron backscattered diffraction. Constitutive equations were developed after considering compensation for strains. The processing maps were established and the optimum processing window was identified. The experimental data and predicted values of flow stresses were in a good agreement with each other. The influence of deformation temperature and strain rates on the microstructure were discussed. The relationship between the recrystallization mechanism and the Zener–Hollomon parameter was investigated as well.

Keywords: Al-Cu-Li-Mg-Ag alloy; constitutive equation; EBSD; recrystallization

---

1. Introduction

Al-Cu-Li alloys have great application potentials in the aerospace field due to their excellent properties, such as low density, high strength, and stiffness [1–3]. The addition of 1 wt% Li in aluminum can approximately reduce its density for 3% and enhance its elastic modulus for 6% [4], which makes Al-Li alloys suitable materials for lower wings and fuselages of airplanes. During the production, hot working on these alloys is generally unavoidable, in which the hot working temperature and strain rate are key parameters for quality control. The process design largely depends on a good understanding of the deformation mechanism. Therefore, it is necessary to investigate systematically the deformation behaviors and microstructure evolution at high temperatures for Al-Cu-Li alloys.

The thermal simulation technique has been an effective method to investigate the behaviors of materials under thermo-mechanical loading. Constitutive models can be developed subsequently to describe the relationship among deformation temperature, strain rate, and flow stress [5]. The flow behaviors of aluminum alloys have been widely simulated by Arrhenius type equations [6–10], which is helpful for process design in a specific deformation condition.

Thermal softening, which was a result of dynamic recovery (DRV) and dynamic recrystallization (DRX), and strain hardening have a close relationship with deformation conditions. An improper choice of deformation parameters may induce surface cracks, inhomogeneous distribution of grains, and flow localization. Therefore, processing maps have been established on the basis of power dissipation to avoid undesirable deformation conditions [11]. Yin [12] developed a processing map for an Al-Cu-Li alloy, where the suitable processing windows were set up at 450–500 °C and 0.01–0.1 s^{-1}. Shen [13] investigated the recrystallization of alloy 2397 at 450–550 °C with strain rates of 0.001–0.1 s^{-1}. It was reported that the dynamical recrystallization (DRX) was controlled by a combination of deformation mechanisms, i.e., grain boundary bulging (BLG) in the original grains and the transformation of low-angle grain boundaries (LAGBs) into high-angle grain boundaries (HAGBs) in the recrystallized grains. Nayan [14] established a wider processing map for alloy 2195 and found DRX was the main softening mechanism in the alloy. Two suitable processing windows are located at temperatures between 400–450 °C with strain rates of 10^{-2}–$10^{-1.5}$ s^{-1} and $10^{-0.5}$–10 s^{-1}. Yu et al. studied the recrystallization in 2A97 alloy and found dynamic precipitation of T_1 phase during hot deformation, and the main recrystallization mechanism was discontinuous dynamic recrystallization (DDRX) [7]. Ou et al.'s [15] study on alloy 2060 indicated that the dominant softening mechanism was DRV and the best hot working window for this alloy was a temperature of 380–500 °C and strain rates of 0.01–3 s^{-1}.

It can be summarized that a different softening mechanism was proposed to explain the hot deformation behavior in Al-Cu-Li alloys, and various process maps were developed for suitable hot working windows. However, systematic investigations on the flow behavior, microstructure evolution, and deformation mechanism of 2060-type alloys are still needed, particularly on the deformation at typical production temperatures, strain rates and large strains. The related recrystallization mechanism is seldom reported as well. Therefore, in this work the hot compression tests on an Al-Cu-Li-Mg-Ag alloy were carried out to acquire the stress–strain data in near practical production conditions. The constitutive models were established and the processing maps were developed to investigate the flow behavior. The microstructure was characterized by using the electron back-scattered diffraction (EBSD) technique, and the relationship between lnZ and the recrystallization mechanism was discussed in detail.

2. Materials and Methods

The nominal composition of the experimental alloy was Al-4.0Cu-0.7Li-0.7Mg-0.4Zn-0.3Mn-0.3Ag-0.11Zr (wt%). The material was semi-continuously cast and homogenized at 410 °C for 4 h followed by 490 °C for 24 h, and then was machined into cylinders with a diameter of 8 mm and a height of 12 mm for hot compression tests.

A Gleeble 3500 thermo-mechanical tester (Dynamic Systems Inc., Austin, TX, USA) was used for the hot compression tests. The test temperatures were between 350 °C and 470 °C and the test strain rate was between 0.01 and 10 s^{-1}. In order to ensure the reliability of data through hot compression tests, graphite was applied between the sample and the die to reduce the interface friction. The specimens were heated up with a heating rate of 2.5 °C/s and held for 3 min at the deformation temperature before the hot compression test to ensure a uniform temperature over the entire specimen, then were compressed to a reduction of 60% in thickness, and finally were quenched into water to reserve the hot deformation microstructures. The microstructures of the test specimens were characterized by a field emission scanning electron microscope (SEM, TESCAN MIRI 3, Brno, Czech Republic) with an electron backscatter diffraction (EBSD) detector (EDAX Inc., Oxford Instruments, Abingdon, UK).

The deformed samples were cut parallel to the compression direction. The examined surface was ground with different sizes of SiC papers and then electro-polished in an electrolyte solution comprised of 10% perchloric acid and 90% alcohol at 20 V for 15 s before observation. After exporting the data in the format cpr from AZtec software (version 3.2, EDAX Inc., Abingdon, UK), EBSD maps were obtained by using HKL Channel 5.

3. Results and Discussion

3.1. Original Microstructure

Figure 1 shows the EBSD orientation map of the Al-Cu-Li-Mg-Ag alloy after homogenization. The grain size is about 190 μm. In the map, the high angle grain boundaries (HAGBs) are depicted in black lines whereas the low angle grain boundaries (LAGBs) are depicted in white lines. HAGBs and LAGBs are differentiated by the misorientation angle, HAGBs are defined as boundaries with misorientation angles higher than 15°, while LAGBs are defined as misorientation angles lower than 15°. As shown in Figure 1b, HAGBs occupy about 70% of the boundaries and the average misorientation angle is around 30°, so the majority of grain boundaries in the starting material are HAGBs.

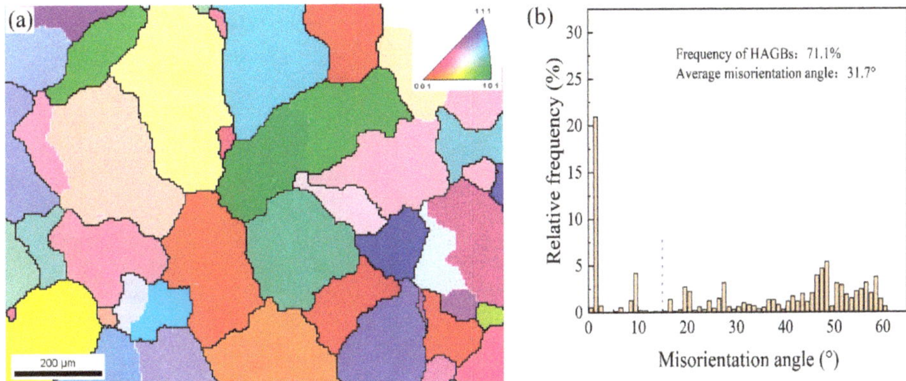

Figure 1. (**a**) Electron back-scattered diffraction (EBSD) map and (**b**) grain boundary misorientation angle of homogenized Al-Cu-Li-Mg-Ag alloy.

3.2. Flow Behavior

Figure 2 shows the true stress–true strain curves of the Al-Cu-Li-Mg-Ag alloy during hot compression. In all compression tests, the flow stress decreases progressively with the increase of strain after a peak stress. This is dynamic softening, which is a result of dynamic recovery and recrystallization [16]. At a given strain rate, the flow stress decreases with increasing deformation temperatures, because of a higher softening effect at higher temperatures, and lower resistance to the motion of dislocations. When the sample is compressed at the same temperature, the peak stress decreases with the decreasing strain rates, as a low strain rate extends the deformation time and offers more time for dynamic softening [17]. However, adiabatic heating due to rapid plastic deformation will lead to the flow softening, especially at higher strain rates [18]. Because of this effect, the corrected flow stress shown in Figure 2c,d has been applied to the data used in following sections.

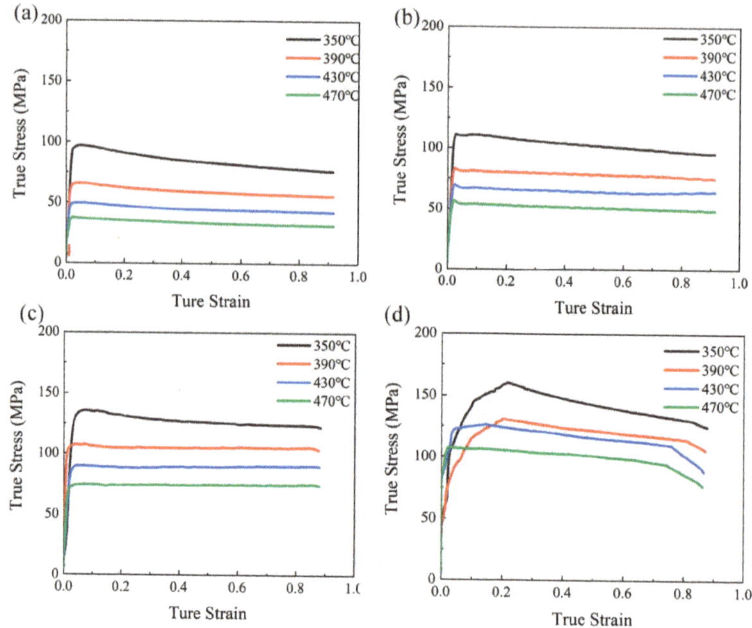

Figure 2. True stress–true strain curves of the Al-Cu-Li-Mg-Ag alloy at various strain rates: (**a**) 0.01 s^{-1}; (**b**) 0.1 s^{-1}; (**c**) 1 s^{-1}; (**d**) 10 s^{-1}.

3.3. Constitutive Equation

The constitutive equation for Al-Cu-Li-Mg-Ag alloy can be developed from the experimental strain–stress curves in Figure 2. In order to understand the influence of deformation temperature and strain rate on the experimental strain–stress curve, the Zener–Hollomon parameter Z is often used [8]:

$$Z = \dot{\varepsilon}\exp(Q/RT) \tag{1}$$

where $\dot{\varepsilon}$ represents the strain rate, Q is the activation energy of hot deformation (kJ/mol), R is the universal gas constant (8.314 kJ mol^{-1}), and T is the deformation temperature. The Z parameter is an important parameter for hot deformation, especially in understanding the microstructure evolution during hot deformation [19].

The relation among stress, strain rate, and temperature during hot deformation can also be expressed by [2]:

$$\dot{\varepsilon} = f(\sigma)\exp(-Q/RT) \tag{2}$$

where $f(\sigma)$ is a function of stress σ and has different expressions at different stress levels. It is generally expressed using a hyperbolic sine function, as

$$f(\sigma) = A\left[\sinh(\alpha\sigma)^{n}\right] \tag{3}$$

where A, α, and n are material constants and $\alpha = \beta/n_1$. n_1 and β can be calculated by the mean slope of $\ln\dot{\varepsilon} - \ln\sigma$ and $\ln\dot{\varepsilon} - \sigma$, respectively, so that the value of α can be obtained thereafter.

From Equations (1)–(3), it can be derived that the values of Q and lnZ are

$$Q = R\left[\frac{\partial\ln\dot{\varepsilon}}{\partial\ln[\sinh(\alpha\sigma)]}\right]_T \left[\frac{\partial\ln[\sinh(\alpha\sigma)]}{\partial(1/T)}\right]_{\dot{\varepsilon}} = RnS \tag{4}$$

$$\ln Z = \ln A + n\ln[\sin h(\alpha\sigma)] \tag{5}$$

where n is defined as the relative change of $\ln\dot{\varepsilon}$ vs. $\ln[\sinh(\alpha\sigma)]$ at constant temperatures and the value of S is defined as the relative change of $\ln[\sinh(\alpha\sigma)]$ vs. 1/T at constant strain rates. Therefore, Q and lnZ can be calculated according to Equations (4) and (5) respectively.

The method to determine the material constants listed above is shown hereafter, using the strain at 0.8 as an example. As a result of linear fitting in Figure 3, the average value of n_1 and β are 8.053 and 0.1004 MPa^{-1}, respectively. Therefore, α can be obtained by β/n_1, which is 0.0125 MPa^{-1}. n is 5.886, and Q is 159.9 kJ mol^{-1}. The value of lnZ can then be figured out on the basis of Equation (1) and the results are listed in Table 1.

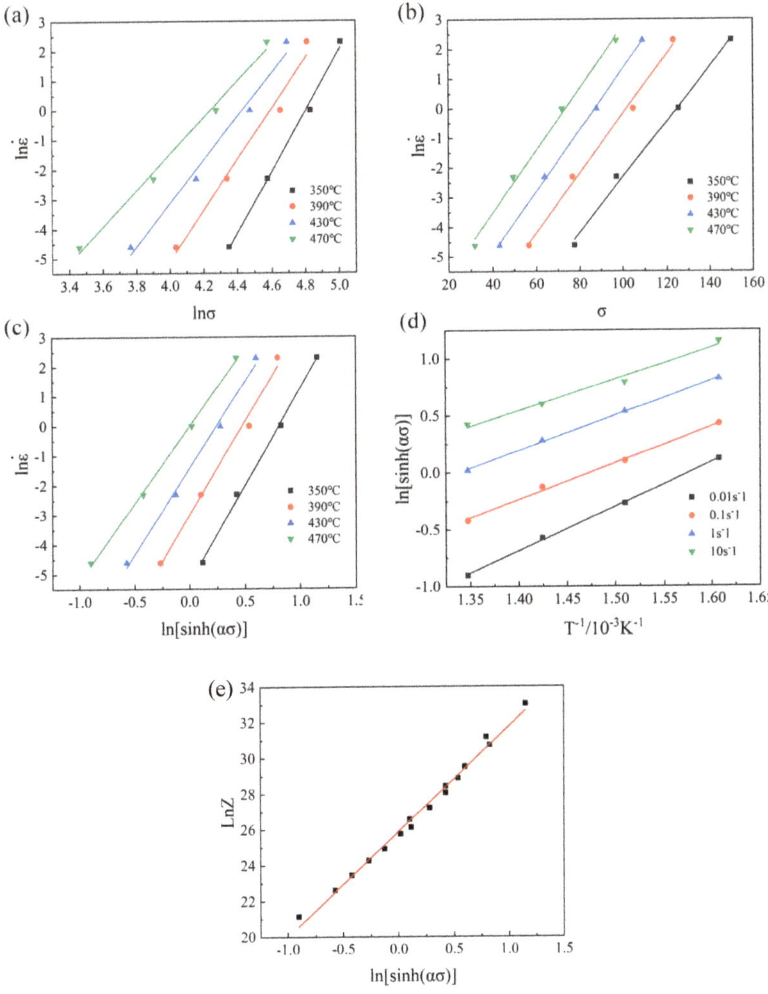

Figure 3. The (**a**) $\ln\dot{\varepsilon}$ vs. $\ln\sigma$, (**b**) $\ln\dot{\varepsilon}$ vs. σ, (**c**) $\ln\dot{\varepsilon}$ vs. $\ln[\sinh(\alpha\sigma)]$; (**d**) $\ln[\sinh(\alpha\sigma)]$ vs. and (**e**) lnZ vs. $\ln[\sinh(\alpha\sigma)]$ curves at stain of 0.8 for the experimental alloy.

Table 1. lnZ at a strain of 0.8 for different deformation condition.

Strain Rate/s^{-1}	350 °C	390 °C	430 °C	470 °C
0.01	26.116	28.419	30.722	33.024
0.1	24.261	26.563	28.866	31.168
1	22.616	24.919	27.221	29.524
10	21.149	23.452	25.754	28.057

lnA is the value of lnZ at ln[sinh(ασ)] = 0 and can be obtained from Figure 3e. Therefore, the constitutive equation for the Al-Cu-Li-Mg-Ag alloy at the strain of 0.8 is:

$$\dot{\varepsilon} = 1.74 \times 10^{11}[\sinh(0.0125\sigma)]^{5.886} \times \exp(-159.9/RT) \tag{6}$$

For a more general expression, different levels of strains should be considered, as it plays an important role in the flow behavior of aluminum alloy (shown in Figure 2). Based on the method of compensation for strain [20], the constitutive equation predicting flow stresses for the Al-Cu-Li-Mg-Ag alloy can be described as:

$$\sigma = \frac{1}{\alpha}\left\{\left(\frac{Z}{A}\right)^{\frac{1}{n}} + \left[\left(\frac{Z}{A}\right)^{\frac{2}{n}} + 1\right]^{\frac{1}{2}}\right\}$$

$$Z = \dot{\varepsilon}\,\exp\left(\frac{Q}{RT}\right) \tag{7}$$

The material constants (α, Q, n, and lnA) in the above equation at various strains ranging from 0.1 to 0.8 with an interval of 0.1 can be calculated in the similar method. The results are shown in Figure 4. It is clear that these constants are strain dependent and can be correlated by the seventh-order polynomial fitting curves to describe accurately the influence of strains.

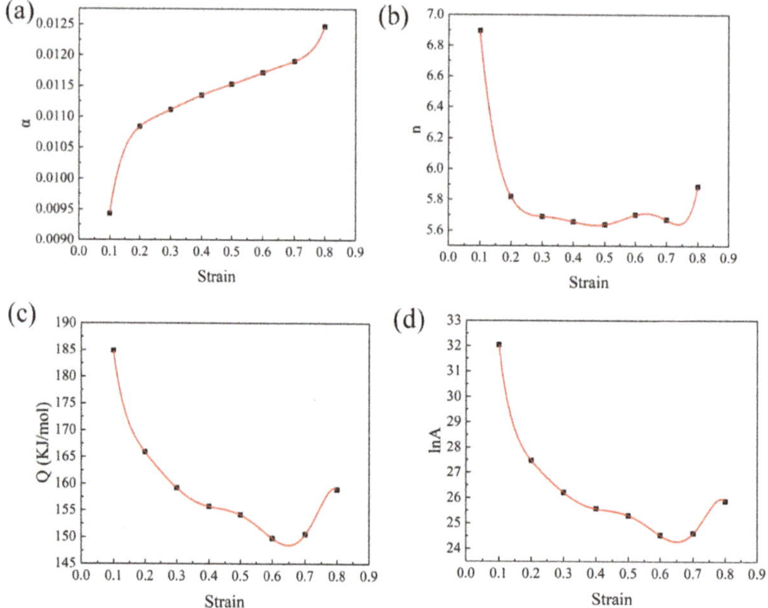

Figure 4. The dependence of material constants (**a**) α; (**b**) n; (**c**) Q;(**d**) lnA on strain.

The predicted values of flow stresses and the experimental data are plotted in Figure 5. The correlation coefficient (R) can be introduced to evaluate the accuracy of the constitutive equations. By fitting the predicted data with the experimental values of flow stresses, the value of R can be figured out as 0.985 with an average relative error (AARE) of 5.4%, which indicates the developed model can predict the high temperature behavior of the experimental Al-Cu-Li-Mg-Ag alloy very well.

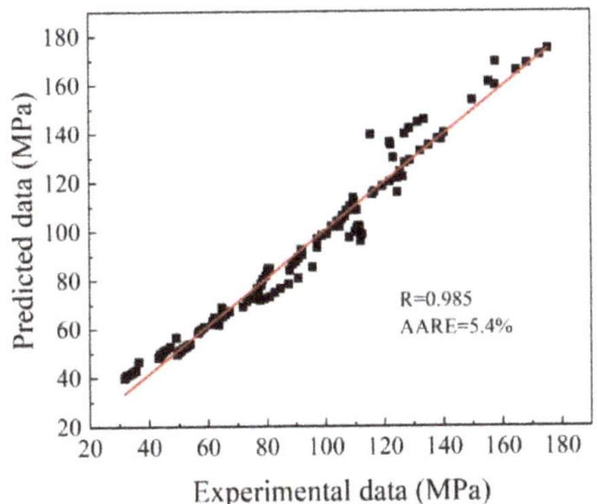

Figure 5. The predicted flow stress vs. the experimental flow stress.

3.4. Hot Processing Map

Hot processing maps can be generated to determine the appropriate process domains for hot working. The values of lg (true stress) versus lg (strain rate) in a compression temperature range between 350 °C and 470 °C are presented in Figure 6a, which are fitted by a cubic spline function. The values of power dissipation (η) are calculated as a function of temperature and strain rate, which are shown as contour lines in Figure 6b and can be used to assess the relationship between power dissipation capacity and microstructural evolution of materials [5]. A dimensionless parameter ξ is also introduced to identify the dependence of flow instability on the temperature and strain rate, and the shaded area in the processing map indicates the unstable domain ($\xi < 0$, Figure 6b).

If a material is deformed within the unstable window, the material is easy to form undesirable microstructures such as adiabatic shear band and intergranular cracking. Therefore, these instability domains ($\xi < 0$) should be avoided when selecting proper parameters for hot working [21,22], while the high power dissipation areas are usually an appropriate processing zone. Based on the processing map, it can be figured out that the optimum process window for this Al-Cu-Li-Mg-Ag alloy locates in the domain at a temperature between 440 °C and 470 °C, and a strain rate between 0.01 s^{-1} and 0.03 s^{-1}. The instable domain is mainly in the window at a temperature range between 350 °C and 430 °C, and a strain rate between 1 s^{-1} and 10 s^{-1}, which is larger than the aluminum alloy 2060 reported, due to greater strains applied in this work [15].

For an accuracy evaluation of processing parameters, the processing map alone is not enough, due to the prediction error of the dynamic materials model [23]. The processing map, composed of power dissipation efficiency and instability domains, connects the microstructure change and the deformation conditions [24,25]. Therefore, microstructure observations are also needed for the purpose of deciding the optimal hot working conditions. Therefore, the microstructures of samples compressed in different domains of the processing map, i.e., the safe domains with high power dissipation (domain A) and low power dissipation (domain B), and unstable domain (domain C) (Figure 6b), will be discussed. The deformation mechanisms will be discussed accordingly.

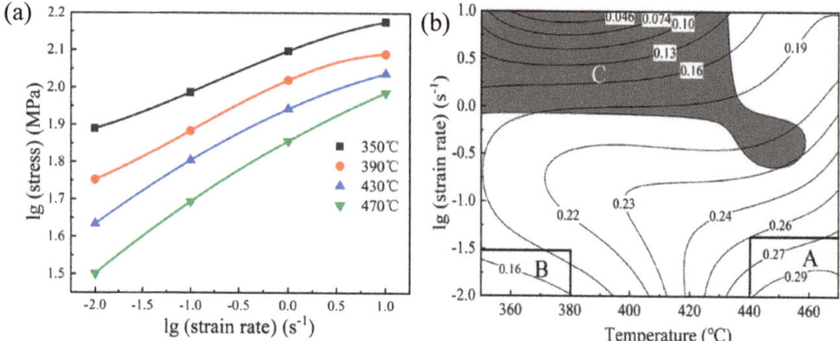

Figure 6. (**a**) The relationship between stress and strain rate, (**b**) processing map at a true strain of 0.8.

3.5. Microstructure Evolution

Figure 7 shows the Zener–Hollomon parameter of the Al-Cu-Li-Mg-Ag alloy deformed at different deformation conditions with a true strain of 0.8. It is obvious that the value of lnZ decreases with the increase of temperature and decrease of strain rate. The highest lnZ is obtained at the area with large strain rates (unstable domain C). The medium values of lnZ (around 25–27) locates at domain B, where both strain rates and deformation temperature are low. The lowest value of lnZ (about 22) locates in domain A, where the deformation temperature is high and the strain rate is low. The influence of deformation temperature and strain rates on microstructure changes will be discussed in the next section.

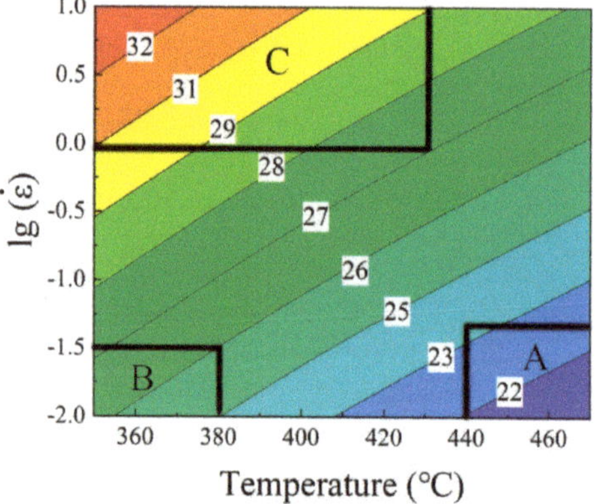

Figure 7. The Zener–Hollomon parameter of the studied alloy at different temperatures and strain rates with a true strain of 0.8. (The values of ln Z are marked on the contour lines; regions A, B, and C correspond the domains shown in Figure 6b).

3.5.1. Effect of Strain Rates

Figure 8 illustrates the effect of strain rate on the microstructure change of the studied Al-Cu-Li-Mg-Ag alloy deformed at 350 °C. The microstructures show typical deformed and recovered structure, i.e., the grains elongate perpendicularly to the compression direction, and many sub-grain boundaries are formed inside the grains. Some recrystallized grains (marked by a black arrow in Figure 8a,b) are observed at lower strain rate (domain A). Adiabatic shear bands, distributed around 45° away from the compression direction as indicated by black rectangles in Figure 8c,d, are observed at higher strain rates (domain C with higher Z value). In this case, severe deformation heat resulted from high strain rates transfer inadequately within limited deformation time, which results in local flow of materials along the plane of the highest shear stress [24,25]. Cracks are expected to generate for the relief of local stress concentrations, thus deformation parameters within domain C are not desirable.

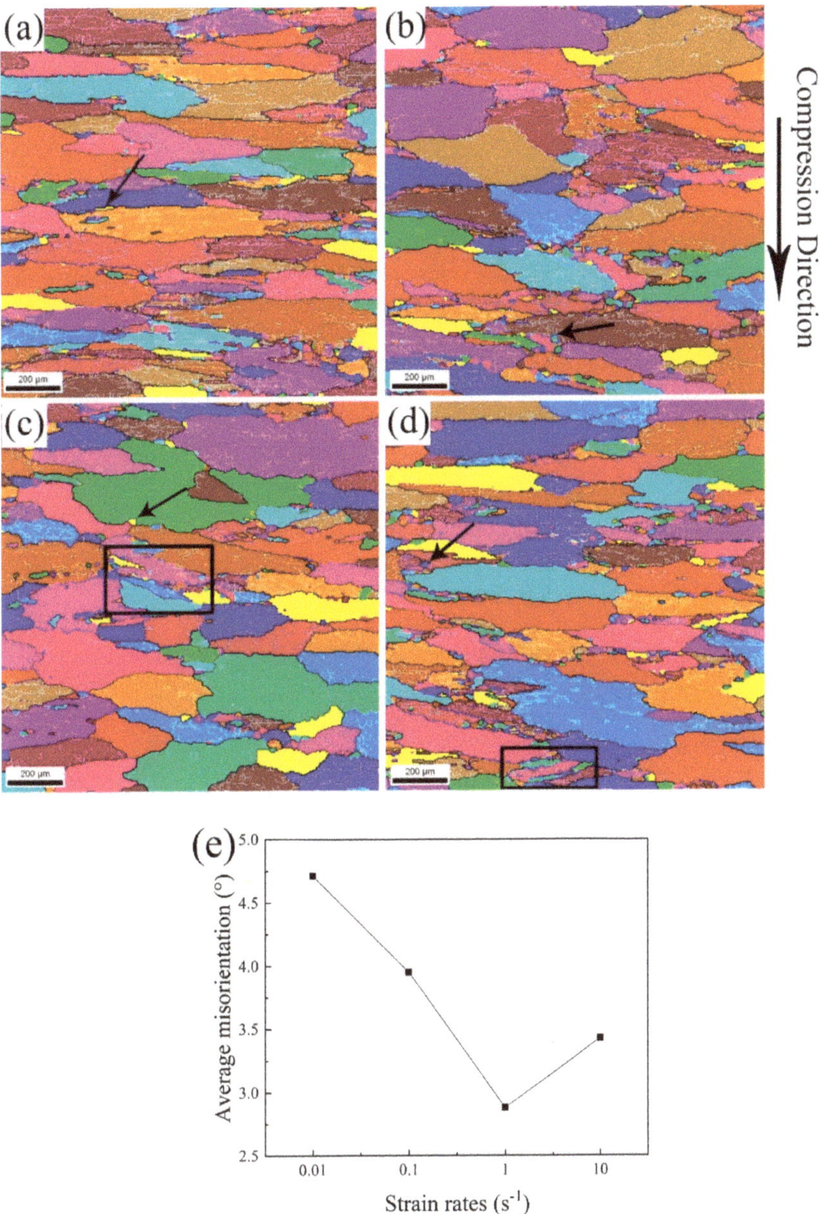

Figure 8. EBSD maps of the Al-Cu-Li-Mg-Ag alloy at 350 °C with various strain rates (**a**) 0.01 s^{-1}; (**b**) 0.1 s^{-1}; (**c**) 1 s^{-1}; (**d**) 10 s^{-1}, and (**e**) average misorientation angle for the four strain rate.

The average misorientation between grains can also be calculated on the basis of EBSD technique. As shown in Figure 8e, the average misorientation decreases firstly with increasing strain rate until strain rates reach 1 s^{-1}, and then increases. Deformation at lower strain rates generally implies that there is adequate time for the dislocation movement, which is evidenced by the massive sub-grain boundaries detected in the grain interior (white lines with misorientation lower than 15° in Figure 8a–c). At higher strain rates (strain rates ≥1 s^{-1}), the adiabatic heating may also be induced, which offers more energy for boundary migration. Thus, the average misorientation values show a non-linear dependence of strain rates.

3.5.2. Effect of Deformation Temperature

Deformation temperature, another factor of Zener–Hollomon parameter Z, has an important influence on the change of microstructures as well. Figure 9 shows the EBSD morphology of the Al-Cu-Li-Mg-Ag alloy deformed at 0.01 s^{-1} and various temperatures. When the sample is deformed at low temperature (350 °C, lnZ = 26.1), dislocation tangling is readily generated, and then dislocation walls develop in the grain interior [5]. So, the microstructure features deformed structures (flat grains) and sub-grains within the grain, as shows in Figure 9a. With increasing temperature, the migration rate of dislocations is enhanced. When the deformation temperature increases to 470 °C (lnZ = 21.1), the improvement of grain boundary mobility promotes the nuclei and growth of recrystallized grains around the original grain boundaries. In addition, dislocation climbs are also enhanced at elevated temperatures, which promote the formation of sub-grain boundaries. Consequently, more recrystallized grains and sub-grains (indicated by white arrows) at the grain boundary and grain interiors (indicated by black arrows) are found than in other lower temperatures (Figure 9a–d). The degree of recrystallization and LAGB% increase with increasing temperature in Figure 9e,f. The fraction of DRX is low regardless of the deformation temperature, which is consistent with the results of Ling [15], which state that the dynamic recovery is the main reason for dynamic softening.

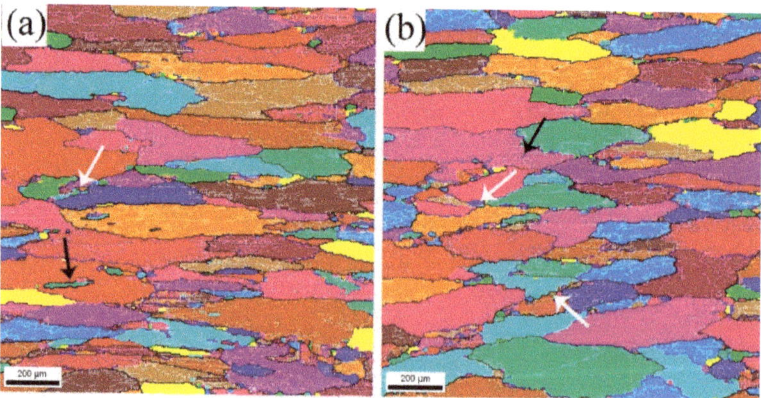

Figure 9. *Cont.*

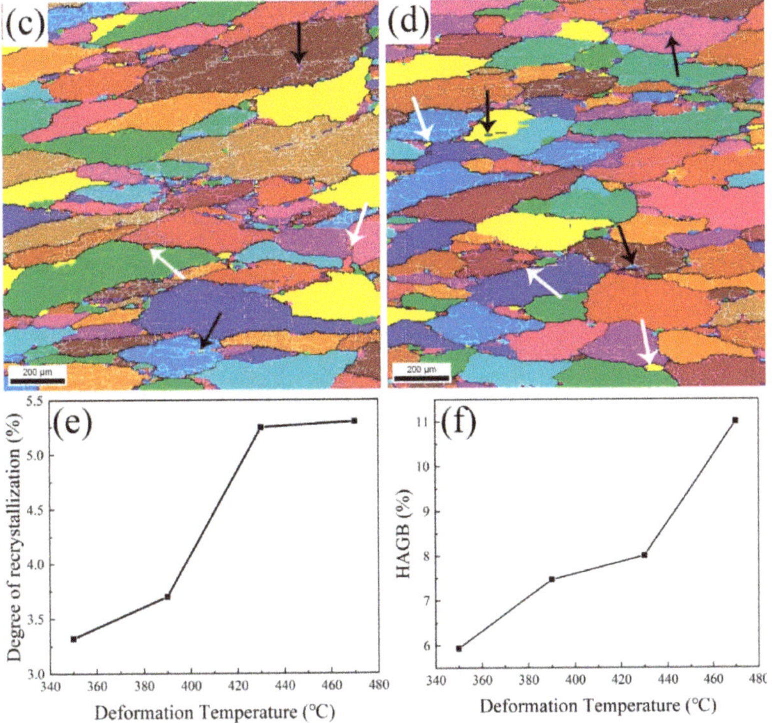

Figure 9. EBSD maps of the Al-Cu-Li-Mg-Ag alloy deformed at a strain rate of $0.01s^{-1}$ and various temperatures (**a**) 350 °C, (**b**) 390 °C, (**c**) 430 °C, (**d**) 470 °C, (**e**) degree of recrystallization and (**f**) the frequency of HAGB.

3.5.3. Recrystallization Mechanism

Based on the processing map (Figure 6b) three representative domains identified by the value of lnZ are chosen to investigate the recrystallization mechanism. Domains A and B are optimum processing regions (A is the highest and B is the lowest value of power dissipation) while domain C is all areas within the unstable domain boundary. Figure 10 illustrates the EBSD morphology of the alloy deformed in different domains: unstable domain C (Figure 10a,b), low power dissipation domain B (Figure 10c), and high power dissipation domain A (Figure 10d). As known to us, the continuous dynamic recrystallization (CDRX) is an important dynamic recrystallization mechanism for high stacking fault energy materials during thermo-mechanical processing. A characteristic of CDRX is the increase of average misorientation angle with the increasing strain [26]. The point-to-origin (cumulative) misorientation is obtained from Figure 10a–d in the grain with similar orientation and shown in Figure 10e. An obvious increase of misorientation gradient is observed inside the grains. Misorientation to the initial grain boundary higher than 15° is observed in all samples, which is strong evidence for CDRX. The fraction of CDRX can be evaluated by the fraction of misorientation angles between 10° and 15° [26–29]. Figure 10f shows the fraction of misorientation angle between 10° and 15° in the four samples, which significant increases compared to the original sample and decreases with decreasing lnZ, suggesting the fraction of CDRX decreases with decreasing lnZ.

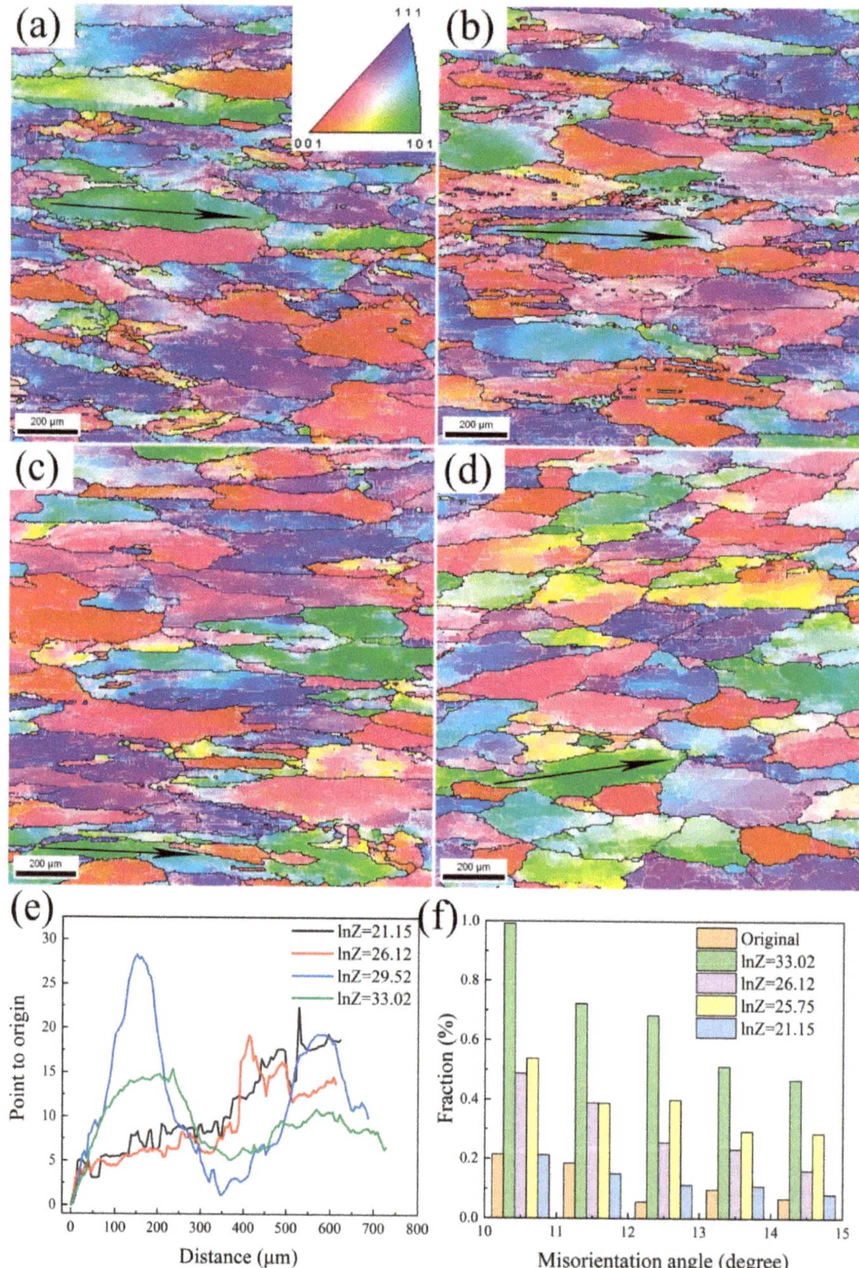

Figure 10. Inverse pole figure maps of the Al-Cu-Li-Mg-Ag alloy under various deformation conditions: (**a**) lnZ = 21.15; (**b**) lnZ = 25.75; (**c**) lnZ = 26.12; (**d**) lnZ = 33.02; (**e**) the changes of misorientation angle along the black lines in (**a**–**d**) and (**f**) the proportion of HAGBs.

Another representative dynamic recrystallization mechanism is discontinuous dynamic recrystallization (DDRX). Grain boundary bulging (BLG), which shows a bow out of grain boundary, is a typical nucleation mechanism of DDRX [30]. The grains with BLG feature, marked by blue arrows, are observed in different deformation conditions, together with the CDRX featured grains indicated by black arrows in Figure 11. The number of grains with BLG features increases with the decreasing of lnZ. Lower values of lnZ usually occur at higher deformation temperatures or lower strain rates, in which case the sub-grain size increases and the dislocation density decreases. Therefore, it can be deduced that the main recrystallization mechanism is gradually transformed from CDRX to DDRX with the decreasing value of lnZ.

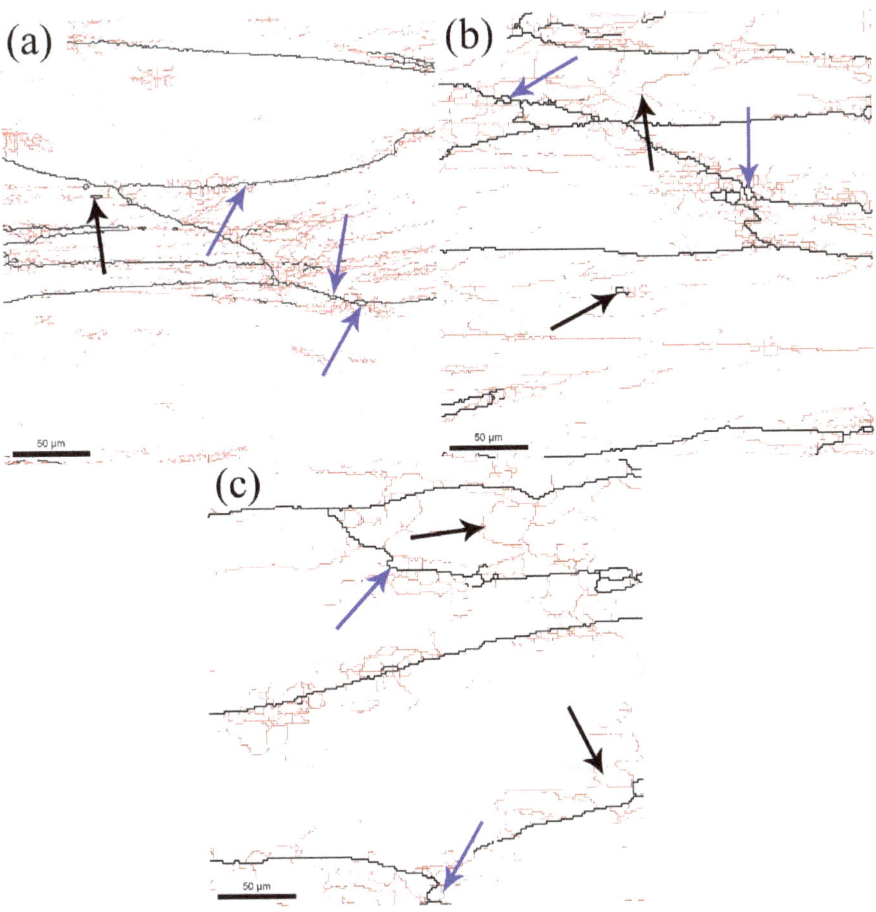

Figure 11. Grain boundary maps of the Al-Cu-Li-Mg-Ag alloy under various lnZ: (**a**) lnZ = 21.15; (**b**) lnZ = 26.12; (**c**) lnZ = 33.02.

Quite a few insoluble constituent particles exist in the Al alloys after homogenization treatments. Most of these insoluble particles are crushed by external force during hot processing and distributed along the processing direction [31]. These particles have a complicated effect on the microstructure evolution during hot processing. On one side, these particles may promote the recrystallization by particle stimulated nucleation. On the other hand, they may inhibit recrystallization by hindering the motion of the grain boundary. The influence of hot deformation parameters on the constituent particles is important and should have significant influence on the microstructure evolution during hot deformation, and should be clarified in the future.

4. Conclusions

In this work, an Al-Cu-Li-Mg-Ag alloy was deformed in a temperature range of 350–470 °C and a strain rate range of 0.01–10 s^{-1}. The flow behavior and microstructure evolution were carefully studied. The conclusions are presented below.

(1) A constitutive equation was developed based on the experimental true stress–strain curves, and for the alloy deformed at a strain of 0.8 it was expressed as

$$\dot{\varepsilon} = 1.74 \times 10^{11} [sinh(0.0125\,\sigma)]^{5.886} \times exp[-159.9/RT]$$

(2) The hot processing map was established and the optimum process window was located in the domain with a temperature between 440 °C and 470 °C, and a strain rate between 0.01 s^{-1} and 0.03 s^{-1}.

(3) Dynamic recovery was the main softening mechanism for the experimental alloy. Dynamic recrystallization was also observed, and CDRX was the main recrystallization mechanism at higher lnZ while DDRX was the main recrystallization mechanism at lower lnZ.

Author Contributions: X.W. and L.C. formulated the experimental design; B.L. and C.L. carried out the Gleeble tests and the microstructure characterization; B.L. analyzed the experimental data and wrote the original draft; X.W. and N.C. discussed the results and made the revisions; L.C. and G.H. applied for the financial support. All authors have read and agreed to the published version of the manuscript.

Funding: The authors acknowledge the financial support by the National Natural Science Foundation of China (51871033, 51421001); the National Defense Basic Scientific Research Program; Key Research and Development Program of China (2016YFB0700403); Venture & Innovation Support Program for Chongqing Overseas Returnees (cx2018002); the Fundamental Research Funds for the Central Universities (2020CDJDCL001); the Open Funding of International Joint Laboratory for Light Alloys (MOE), Chongqing University. X.W. thanks the support of the "111" Project (B16007) by the Ministry of Education and the State Administration of Foreign Experts Affairs of China.

Conflicts of Interest: The authors declare no conflict of interest.

References

1. Alexopoulos, N.D.; Migklis, E.; Stylianos, A.; Myriounis, D.P. Fatigue behavior of the aeronautical Al–Li (2198) aluminum alloy under constant amplitude loading. *Int. J. Fatigue* **2013**, *56*, 95–105. [CrossRef]

2. Li, B.; Pan, Q.; Yin, Z. Characterization of hot deformation behavior of as-homogenized Al–Cu–Li–Sc–Zr alloy using processing maps. *Mater. Sci. Eng. A* **2014**, *614*, 199–206. [CrossRef]

3. Yu, X.X.; Zhang, Y.R.; Yin, D.F.; Yu, Z.M.; Li, S.F. Characterization of hot deformation behavior of a novel Al–Cu–Li alloy using processing maps. *Acta Metall. Sin.* **2015**, *28*, 817–825. [CrossRef]

4. Rioja, R.J.; Liu, J. The evolution of Al-Li base products for aerospace and space applications. *Metall. Mater Trans A* **2012**, *43*, 3325–3337. [CrossRef]

5. Liao, B.; Cao, L.F.; Wu, X.D.; Zou, Y.; Huang, G.J.; Rometsch, P.A.; Couper, M.J.; Liu, Q. Effect of heat treatment condition on the flow behavior and recrystallization mechanisms of aluminum alloy 7055. *Materials* **2019**, *12*, 311. [CrossRef]

6. Zheng, X.; Luo, P.; Chu, Z.; Xu, J.; Wang, F. Plastic flow behavior and microstructure characteristics of light-weight 2060 Al-Li alloy. *Mater. Sci. Eng. A* **2018**, *736*, 465–471. [CrossRef]

7. Yu, W.; Li, H.; Du, R.; You, W.; Zhao, M.; Wang, Z.A. Characteristic constitution model and microstructure of an Al–3.5Cu–1.5Li alloy subjected to thermal deformation. *Mater. Charact.* **2018**, *145*, 53–64. [CrossRef]

8. Zhao, Q.; Chen, W.; Lin, J.; Huang, S.S.; Xia, X.S. Hot deformation behavior of 7A04 aluminum alloy at elevated temperature: Constitutive modeling and verification. *Int. J. Mater. Form.* **2019**, *13*, 293–302.

9. Chen, L.; Zhao, G.; Yu, J.; Zhang, W. Constitutive analysis of homogenized 7005 aluminum alloy at evaluated temperature for extrusion process. *Mater. Des.* **2015**, *66*, 129–136. [CrossRef]

10. Li, J.; Li, F.; Cai, J.; Wang, R.; Yuan, Z.; Xue, F. Flow behavior modeling of the 7050 aluminum alloy at elevated temperatures considering the compensation of strain. *Mater. Des.* **2012**, *42*, 369–377. [CrossRef]

11. Liu, Y.; Geng, C.; Lin, Q.; Xiao, Y.; Xu, J.; Kang, W. Study on hot deformation behavior and intrinsic workability of 6063 aluminum alloys using 3D processing map. *J. Alloys Compd.* **2017**, *713*, 212–221. [CrossRef]

12. Yin, H.; Li, H.; Su, X.; Huang, D. Processing maps and microstructural evolution of isothermal compressed Al–Cu–Li alloy. *Mater. Sci. Eng. A* **2013**, *586*, 115–122. [CrossRef]

13. Shen, B.; Deng, L.; Wang, X. A new dynamic recrystallisation model of an extruded Al–Cu–Li alloy during high-temperature deformation. *Mater. Sci. Eng. A* **2015**, *625*, 288–295. [CrossRef]

14. Nayan, N.; Murty, S.V.S.N.; Chhangani, S.; Prakash, A.; Prasad, M.J.N.V.; Samajdar, I. Effect of temperature and strain rate on hot deformation behavior and microstructure of Al–Cu–Li alloy. *J. Alloys Compd.* **2017**, *723*, 548–558. [CrossRef]

15. Ou, L.; Zheng, Z.; Nie, Y.; Jian, H. Hot deformation behavior of 2060 alloy. *J. Alloys Compd.* **2015**, *648*, 681–689. [CrossRef]

16. Zhang, C.S.; Wang, C.X.; Guo, R.; Zhao, G.Q.; Chen, L.; Sun, W.C.; Wang, X.B. Investigation of dynamic recrystallization and modelling of microstructure evolution of an Al–Mg–Si aluminum alloy during high-temperature deformation. *J. Alloys Compd.* **2019**, *773*, 59–70. [CrossRef]

17. Zhao, J.; Deng, Y.; Tang, J.; Zhang, J. Influence of strain rate on hot deformation behavior and recrystallization behavior under isothermal compression of Al–Zn–Mg–Cu alloy. *J. Alloys Compd.* **2019**, *809*, 151788. [CrossRef]

18. Liu, L.; Wu, Y.; Gong, H.; Dong, F.; Ahmad, A.S. Modified kinetic model for describing continuous dynamic recrystallization behavior of Al 2219 alloy during hot deformation process. *J. Alloys Compd.* **2020**, *817*, 153301. [CrossRef]

19. Liu, S.H.; Pan, Q.L.; Li, H.; Huang, Z.Q.; Li, K.; He, X.; Li, X.Y. Characterization of hot deformation behavior and constitutive modelling of Al–Mg–Si–Mn–Cr alloy. *J. Mater. Sci.* **2019**, *54*, 4366–4383. [CrossRef]

20. Xiang, Y.; Xiao, S.; Tang, Z.; Zhou, Y. The flow behavior of homogenized Al-Mg-Si-La aluminum alloy during hot deformation. *Mater. Res. Express* **2019**, *6*, 066563. [CrossRef]

21. Wang, Y.; Zhao, G.; Xu, X.; Chen, X.; Zhang, C. Constitutive modeling, processing map establishment and microstructure analysis of spray deposited Al-Cu-Li alloy 2195. *J. Alloys Compd.* **2019**, *779*, 735–751. [CrossRef]

22. Hu, H.E.; Wang, X.Y.; Deng, L. High temperature deformation behavior and optimal hot processing parameters of Al–Si eutectic alloy. *Mater. Sci. Eng. A* **2013**, *576*, 45–51. [CrossRef]

23. Jenab, A.; Karimi Taheri, A. Experimental investigation of the hot deformation behavior of AA7075: Development and comparison of flow localization parameter and dynamic material model processing maps. *Int. J. Mech. Sci.* **2014**, *78*, 97–105. [CrossRef]

24. Zhang, F.; Sun, J.L.; Shen, J.; Yan, X.D.; Chen, J. Flow behavior and processing maps of 2099 alloy. *Mater. Sci. Eng. A* **2014**, *613*, 141–147. [CrossRef]

25. Kai, X.; Chen, C.; Sun, X.; Wang, C.; Zhao, Y. Hot deformation behavior and optimization of processing parameters of a typical high-strength Al–Mg–Si alloy. *Mater. Des.* **2016**, *90*, 1151–1158. [CrossRef]

26. Huang, K.; Logé, R.E. A review of dynamic recrystallization phenomena in metallic materials. *Mater. Des.* **2016**, *111*, 548–574. [CrossRef]

27. Chamanfar, A.; Alamoudi, T.M.; Nanninga, E.N.; Misiolek, Z.W. Analysis of flow stress and microstructure during hot compression of 6099 aluminum alloy (AA6099). *Mater. Sci. Eng. A* **2019**, *743*, 684–696. [CrossRef]

28. Mandal, S.; Bhaduri, A.K.; Sarma, V.S. A study on microstructural evolution and dynamic recrystallization during isothermal deformation of a Ti-modified austenitic stainless steel. *Metall. Mater. Trans. A* **2010**, *42*, 1062–1072. [CrossRef]

29. Liao, B.; Wu, X.D.; Cao, L.F.; Huang, G.J.; Wang, Z.A.; Liu, Q. The microstructural evolution of aluminum alloy 7055 manufactured by hot thermo-mechanical process. *J. Alloys Compd.* **2019**, *796*, 103–110. [CrossRef]

30. Shimizu, I. Theories and applicability of grain size piezometers: The role of dynamic recrystallization mechanisms. *J. Struct. Geol.* **2008**, *30*, 899–917. [CrossRef]

31. Mcqueen, H.J.; Spigarelli, S.; Kastner, M.E.; Evangelista, E. *Hot Deformation and Processing of Aluminum Alloys*; CRC Press: Boca Raton, FL, USA, 2011.

Article

Effect of Equal Channel Angular Pressing on Microstructure and Mechanical Properties of a Cu-Mg Alloy

Muzhi Ma [1], Xi Zhang [1], Zhou Li [1,2], Zhu Xiao [1,3,*], Hongyun Jiang [4], Ziqi Xia [1] and Hanyan Huang [1]

[1] School of Materials Science and Engineering, Central South University, Changsha 410083, China; mamuzhimse@163.com (M.M.); zhang_xi@csu.edu.cn (X.Z.); lizhou6931@163.com (Z.L.); xiaziqi@csu.edu.cn (Z.X.); huanghanyan@csu.edu.cn (H.H.)

[2] State Key Laboratory of Powder Metallurgy, Central South University, Changsha 410083, China

[3] Key Laboratory of Non-ferrous Metal Materials Science and Engineering, Ministry of Education, Changsha 410083, China

[4] Zhejiang Tianning Alloy Material Co., Ltd., Jinhua 321002, China; jianghongyun@zjtnhj.com

* Correspondence: xiaozhumse@163.com

Received: 26 April 2020; Accepted: 26 May 2020; Published: 27 May 2020

Abstract: A Cu-0.43Mg (wt.%) alloy was processed by equal channel angular pressing (ECAP) up to eight passes via a processing route (Bc). The hardness distribution on the longitudinal and transverse sections was collected and the microstructure in the central and bottom regions on the longitudinal section was examined. The result showed that the hardness was improved significantly at the initial stage of the ECAP process, and the lower hardness region appeared at the area nearby the bottom surface. With the number of ECAP passes, the hardness gently increased and finally became saturated. The inhomogeneity of the hardness distribution along the normal direction gradually weakened and finally disappeared. The shear microstructure in the central region was different from that in the bottom region after one ECAP pass, and they became similar to each other after two ECAP passes, except the rotation angle of the elongated grains. With the further increasing ECAP passes, there was no obvious microstructure difference between the central and bottom regions. The inhomogeneities of the hardness and the microstructure along the normal direction in the alloy after one ECAP passes should be attributed to the non-zero outer arc of curvature of the ECAP die and the friction between the bottom surface of the billets and the ECAP die walls. The yield strength of the alloy increased from 124 MPa before the ECAP process to 555 MPa after eight ECAP passes. The improvement of yield strengths of the ECAPed Cu-Mg alloy should be mainly attributed to the dislocation strengthening and the grain boundary strengthening.

Keywords: Cu-Mg alloy; equal channel angular pressing (ECAP); microstructure; mechanical properties

---

1. Introduction

Copper alloys play a vital role in mechanical manufacturing, railway transportation, electrical and electronic industries [1–4]. Cu-Mg alloys have been widely applied as the contact wire and other conductor materials in the high-speed rail industry due to their good combination of high strength and high electrical conductivity [5–8]. However, the present Cu-Mg alloy contact wire is currently unable to meet the new requirements of higher speed, heavier haul and lower power loss of high-speed railway vehicles [9]. Many researchers have attempted to improve the comprehensive properties of the Cu-Mg alloy via the microalloying method [10–13], but the solute atoms introduced by the microalloying method could significantly and inevitably decrease the electrical conductivity of the Cu-Mg alloy [14].

The severe plastic deformation (SPD) technique can effectively improve the strength of copper alloy without decreasing its high electrical conductivity [15]. The most attractive one is the equal channel angular pressing (ECAP) technique. It is an attractive SPD method to produce copper alloy with an ultra-fine grain (UFG). The ECAP process involves that the metallic material is deformed by pressing through an equal cross-sectional channel with the right or obtuse angle [16]. After multiple ECAP processes, the micro- or nano-scaled UFGed microstructure can be obtained, leading to the high strength of alloys according to the Hall–Petch relationship [17–20]. The dislocations and grain boundaries introduced by the ECAP process have a limited influence on the electron scattering, and thus the good electrical conductivity of the alloy can be maintained [14]. The ECAP process combined with the continuous forming (CONFORM) process can be easily incorporated to produce Cu-Mg alloy contact wire with a standard length of more than 1400 m [21]. However, the ECAP process often results in an inhomogeneous shear strain in the material [22], and leads to the different microstructure and properties in different regions of the ECAPed material [23,24]. Taking pure Al as an example, after one ECAP pass, the hardness in the region adjacent to the bottom surface (30–35 HV) was much lower than those in the other regions (40–50 HV) [23,24]. This inhomogeneity phenomenon is undoubtedly harmful to the comprehensive properties of ECAPed materials, and the microstructure evolution and its effect on the mechanical properties of the highly conductive copper alloys are worth investigating.

In this study, the effect of ECAP on the microstructure and mechanical properties of a Cu-0.43Mg (wt.%) alloy was investigated in detail. The hardness on the longitudinal and transverse sections was measured and analyzed. The microstructure in the central and bottom regions on the longitudinal section was observed by the electron back-scattered diffraction (EBSD) technique. The mechanism of the inhomogeneity phenomenon in the alloy during the ECAP process was discussed. The strengthening mechanisms of the ECAPed Cu-Mg alloy were also analyzed quantitatively.

2. Materials and Methods

The Cu-0.43Mg (wt.%) alloy was melted with the raw materials of pure copper and pure magnesium using a medium frequency induction furnace, and its chemical composition was measured using a SPECTROBLUE inductively coupled plasma optical emission spectrometer (ICP-OES) (SPECTRO Analytical Instruments GmbH, Kleve, Düsseldorf, Nordrhein-Westfalen, Germany). The ingot was homogenization treated at 780 °C for 1 h, and then hot-rolled from 35 to 12 mm in thickness, followed by water quenching. The cylindrical billets for the ECAP test were cut from the hot-rolled plate with a dimension of Φ 10 × ~70 mm. The ECAP process was performed at room temperature, on a 150T four-column hydraulic press, with the plug speed of ~10 mm/s. Figure 1a shows the schematic illustration of the ECAP process. The ECAP die was a two-piece split die, with the inner channel angle (Φ) of 110° and the outer arc of curvature (Ψ) of 30°. A suspension containing graphite and machine oil was used as the lubricant to reduce the friction between the billets and the ECAP die wall. The billets were repetitively ECAP-processed up to 8 passes via a processing route (Bc), where the billet was rotated by 90° about the longitudinal axis in the same direction between each consecutive pass [16]. The laboratory coordinate system of the ECAP process was defined as follows: X, Y and Z axes were parallel to the extrusion direction (ED), normal direction (ND) and transverse direction (TD), respectively [25].

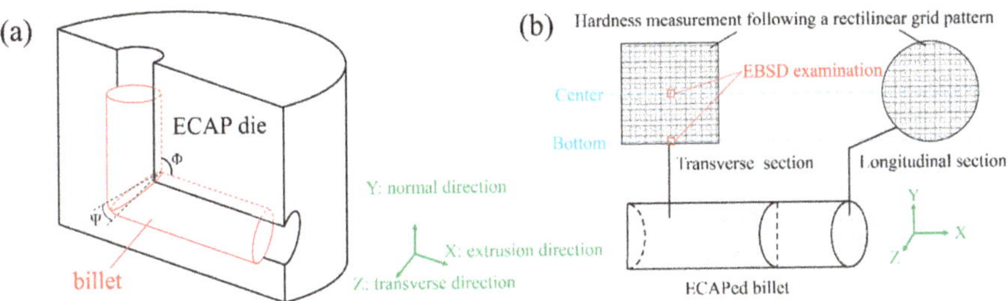

Figure 1. Schematic illustrations of (**a**) the equal channel angular pressing (ECAP) process and (**b**) the hardness measurement and electron back-scattered diffraction (EBSD) examination.

Figure 1b shows the schematic illustration of the hardness measurement and EBSD examination. Vickers hardness on the longitudinal and transverse sections of the ECAPed billets was measured following a rectilinear grid pattern with a spacing of 0.5 mm using a HV-5 microhardness tester, with a load of 1 kg and a dwell time of 15 s. The hardness was visualized in the form of a color-coded contour map in order to intuitively show the hardness distribution. The microstructure in the central and bottom regions on the longitudinal section was examined by the EBSD technique using an FEI Helios Nanolab 600i scanning electron microscope (SEM) equipped with a NordlysMax2 EBSD detector operating at an accelerating voltage of 20 keV and an electric current of 1 nA. In this case, the copper specimen had a lateral resolution of ~40 nm and a longitudinal resolution of ~100 nm. The EBSD data were analyzed using the HKL Channel5 software. The average grain size was characterized using boundaries with the misorientation angles larger than 15° and allowing completion down to 10°, and the grain areas were converted to the equivalent diameters. For the mean boundary spacing, the boundary was detected according to the critical misorientation angle of 2°, and then the mean boundary spacing was measured using the line intercept method. In addition, the boundaries with the misorientation angles larger than 15° were defined as the high-angle boundaries (HABs), and the boundaries with the misorientation angles lower than 15° but larger than 2° were defined as the low-angle boundaries (LABs).

3. Results

Figure 2 shows the color-coded contour maps of the hardness on the longitudinal section of the hot-rolled and ECAPed Cu-Mg alloy. The hardness of the hot-rolled Cu-Mg alloy was very low (75 ± 1 HV) but the distribution of hardness was homogeneous (Figure 2a). After one ECAP pass, the hardness nearly doubled to ~140±3 HV (Figure 2b). However, the hardness adjacent to the bottom surface (121 ± 4 HV) was obviously lower than that in the other regions (141 ± 2 HV), indicating an inhomogeneous hardness distribution along the Y direction. After two ECAP passes, the hardness increased to ~155 ± 3 HV, except the low-hardness region adjacent to the bottom surface (145 ± 3 HV) (Figure 2c). In addition, the degree of hardness inhomogeneity weakened compared with that after one ECAP pass. After four, six and eight ECAP passes, the hardness gently increased to 168 ± 3 HV, 177 ± 2 HV and 181 ± 2 HV, respectively, and the hardness inhomogeneity along the Y direction disappeared (Figure 2d–f). Figure 3 shows the color-coded contour maps of the hardness on the transverse section of the hot-rolled and ECAPed Cu-Mg alloy. The result was similar to that on the longitudinal section. On the one hand, the hardness was improved significantly from 75 ± 1 HV before the ECAP process to ~140 ± 3 HV at the initial stage of the ECAP process. Then, it gently increased to ~155 ± 3 HV after two ECAP passes, 167 ± 3 HV after four ECAP passes, 177 ± 2 HV after six ECAP passes and finally achieved stability at 181 ± 2 HV after eight ECAP passes. On the other hand, the lower-hardness regions with the values of 121 ± 2 HV and 144 ± 3 HV occurred at the area near the bottom surface after one and two ECAP passes, respectively. Then, the degree of hardness inhomogeneity along the Y direction gradually weakened and finally disappeared.

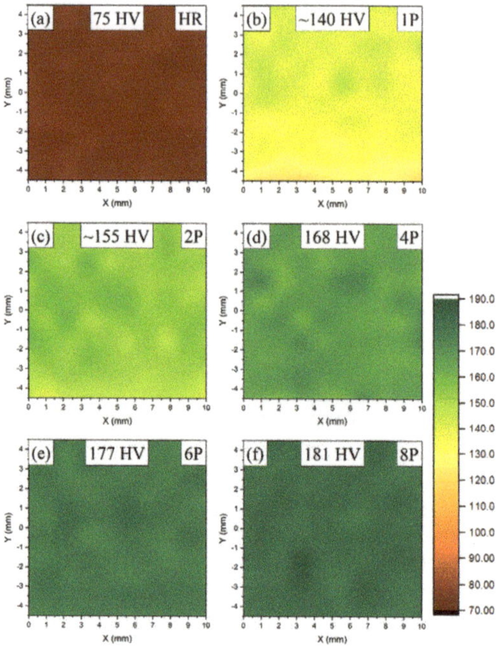

Figure 2. Color-coded contour maps of the hardness on the longitudinal section of the Cu-Mg alloy (**a**) before the ECAP process and after (**b**) 1, (**c**) 2, (**d**) 4, (**e**) 6 and (**f**) 8 ECAP passes.

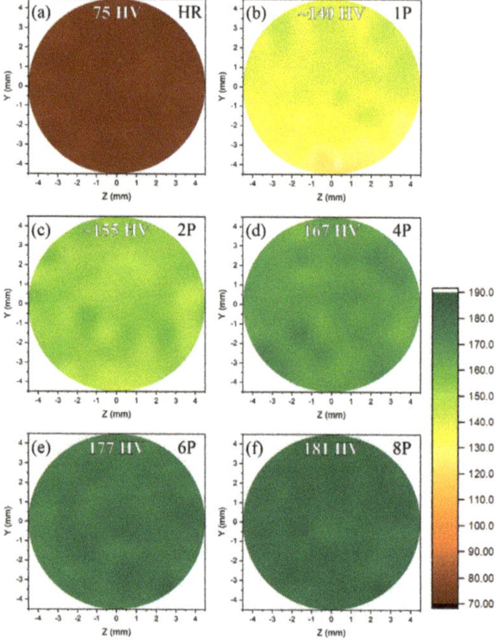

Figure 3. Color-coded contour maps of the hardness on the transverse section of the Cu-Mg alloy (**a**) before the ECAP process and after (**b**) 1, (**c**) 2, (**d**) 4, (**e**) 6 and (**f**) 8 ECAP passes.

Figures 4 and 5 show the inverse pole figure (IPF) coloring maps in the central and bottom regions on the longitudinal section of the ECAPed Cu-Mg alloy, respectively. The direction label of the ECAP process and the coloring triangle of the cupreous crystallographic orientation are shown on the lower-right regions of Figures 4 and 5. After one ECAP pass, the central region showed the shear microstructure, where the grains were elongated along the direction counter-clockwise-rotated ~45° with the ED (Figure 4a). The drastic color change inside the elongated grains indicated that a number of LABs subdivided the original hot-rolled grains. However, the bottom region showed a different microstructure consisting of the distorted equiaxed grains and twins (Figure 5a). The grain morphology was similar to that before the ECAP process, even though some dislocations appeared inside the grains. After two ECAP passes, the microstructures in the central and bottom regions were both composed of the elongated grains, but the rotation angle of the elongated grains in the central region (~45°) was higher than that in the bottom region (~20–30°) (Figures 4b and 5b). After four and six ECAP passes, the majority of the elongated grains in the central and bottom regions was refined and the grain sizes showed bimodal distributions (Figure 4c,d and Figure 5c,d). After eight ECAP passes, the refined grains were surrounded by the HABs, indicating that the two regions achieved the UFGed microstructure (Figures 4e and 5e).

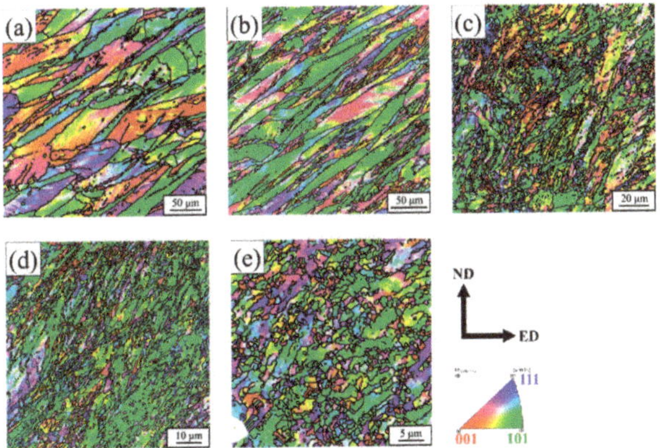

Figure 4. Inverse pole figure (IPF) coloring maps in the central region on the longitudinal section of the ECAPed Cu-Mg alloy: (**a**) 1, (**b**) 2, (**c**) 4, (**d**) 6 and (**e**) 8 ECAP passes.

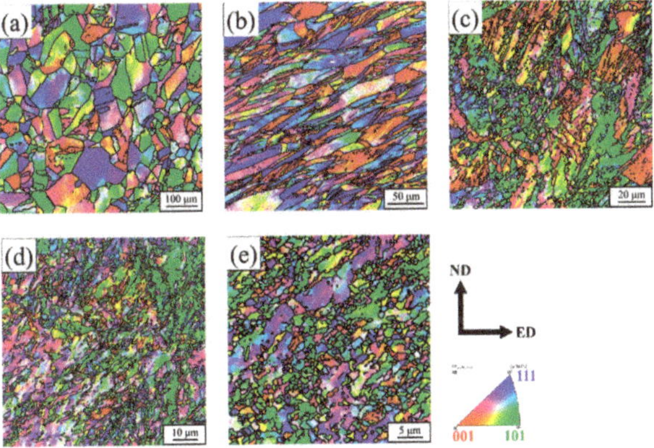

Figure 5. IPF coloring maps in the bottom region on the longitudinal section of the ECAPed Cu-Mg alloy: (**a**) 1, (**b**) 2, (**c**) 4, (**d**) 6 and (**e**) 8 ECAP passes.

Figure 6 shows the microstructure information in the central and bottom regions on the longitudinal section of the ECAPed Cu-Mg alloy. The average grain size and the mean boundary spacing decreased with the number of ECAP passes (Figure 6a,b). After the first four ECAP passes, the average grain size decreased rapidly from dozens of micrometers before the ECAP process [15] to two or three micrometers, and the mean boundary spacing also decreased sharply from several hundred micrometers before the ECAP process [15] to less than one micrometer. In the subsequent ECAP process, the average grain size and the mean boundary spacing gradually reached saturation. In addition, it should be noted that at the initial stage of the ECAP process, the two parameters in the central region were lower than those in the bottom region, but this difference disappeared after the fourth ECAP pass. For the other two parameters, at the initial stage of the ECAP process, the dislocations and the subgrain boundaries were predominant, and thus the fraction of HABs and the mean misorientation angle of LABs showed the low values (Figure 6c,d). With of the increasing ECAP passes, the subgrain boundaries absorbed dislocations to increase its misorientation angle and transform into HABs, leading to the increase in the fraction of HABs and the mean misorientation angle of LABs.

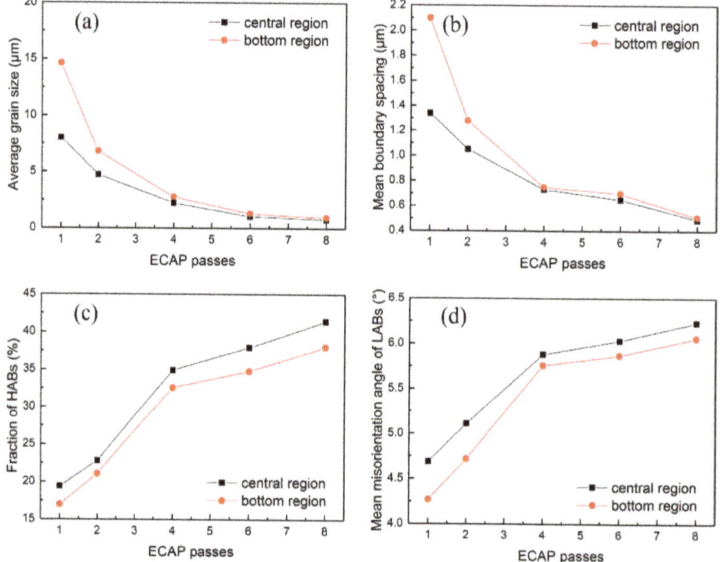

Figure 6. Microstructure information in the central and bottom regions on the longitudinal section of the ECAPed Cu-Mg alloy: (**a**) average grain size, (**b**) mean boundary spacing, (**c**) fraction of high-angle boundaries (HABs) and (**d**) mean misorientation angle of low-angle boundaries (LABs).

4. Discussion

4.1. Mechanism for Inhomogeneity Phenomenon

It is noted that the hardness and microstructure of the Cu-Mg alloy after one ECAP pass are very inhomogeneous along the Y direction, however, this inhomogeneity phenomenon gradually weakens and finally disappears with the number of ECAP passes. This is consistent with some previous research works on pure Al and Al alloys [26–28]. The inhomogeneity phenomenon at the initial stage of the ECAP process could be attributed to the formation of a corner gap [26–28]. At the corner gap, the billets no longer contact with the ECAP die walls and in turn the die walls no longer impose the shear strain on the billets. Thus, the inhomogeneous strain imposed on the billets results in the uneven distribution of the hardness and microstructure. However, the formation of a corner gap is related to the outer arc of curvature of the ECAP die. It was reported that a smaller corner gap was found when using an ECAP die with a high outer arc of curvature [29]. Therefore, the reason for the inhomogeneity phenomenon should be reconsidered.

An ideal ECAP process is considered as a simple shear deformation, with the shear direction from the inner corner to the outer corner of channel and the shear plane located on the intersection plane of the two parts of the channel [22]. However, the inevitable shear stress derived from the friction between the bottom surface of the billets and the ECAP die walls is often ignored [30–32]. Thus, the practical ECAP process becomes a complicated deformation combining the ideal simple shear deformation and the frictional shear deformation near the bottom surface. In addition, the non-zero outer arc of curvature expands the main plastic deformation zone (MPDZ) of the ideal ECAP process from the intersection plane of the two parts of the channel into the fan-shaped zone, which is sharp at the inner corner and broadens towards the outer corner [33,34]. When the billets pass through the MPDZ, the distribution of the equivalent strain rate ($\dot{\varepsilon}$) along the Y direction becomes inhomogeneous [35]. The more adjacent to the bottom surface the region is, the lower the equivalent strain rate is. For a special situation of $\dot{\varepsilon} = 0$, it means that only rigid-body rotation occurs in the grains, rather than plastic deformation.

These additional factors have a great influence on the microstructure and mechanical properties of the ECAPed Cu-Mg alloy. Firstly, in the initial stage of the ECAP process, the region adjacent to the bottom surface goes through such little plastic deformation that only a small number of dislocations need be activated to accommodate it. When the billet is pressed through the ECAP die, only the kinematical rigid-body rotation occurs in the alloy [36]. Thus, the grain morphology could not obviously change compared with the hot-rolled one, except a slight distortion of grains, and the hardness is also lower than those in other regions. Secondly, the region adjacent to the bottom surface shows the lower rotation angle of the elongated grains due to the additional frictional shear stress parallel to the ED. The EBSD result confirmed that the rotation angle of the elongated grains in the bottom region after two ECAP passes is 20°–30° but that in the central region, it is ~45°. Thirdly, the grain refinement during the ECAP process is based on the dislocation subdivision mechanism [37]. The accumulation rate of dislocations determines the rate of microstructure evolution. Therefore, in the bottom region, the average grain size and the mean boundary spacing are higher, and the fraction of HABs and the mean misorientation angle of LABs are lower. In addition, it should be noted that the route Bc also plays an important role in the weakening and disappearance of inhomogeneity. The bottom region after the first ECAP pass is not the bottom region after the second ECAP pass.

4.2. Strengthening Mechanisms and Strengths Calculation

The strengthening mechanisms of the ECAPed Cu-Mg alloy can be considered from four aspects and its yield strength can be expressed as Equation (1) [38]:

$$\sigma_y = \sigma_{LF} + \sigma_{SS} + \sigma_D + \sigma_{GB} \tag{1}$$

where σ_y is the yield strength of the ECAPed Cu-Mg alloy, σ_{LF} is the lattice friction or the Peierls–Nabarro stress, σ_{SS} is the strength contribution from the solid solution strengthening, σ_D is the strength contribution from the dislocation strengthening and σ_{GB} is the strength contribution from the grain boundary strengthening. A previous research showed that σ_{LF} and σ_{SS} of the ECAPed Cu-Mg alloy had little relation with the number of ECAP passes, with the values of 68.3 and 48.7 MPa, respectively [15].

The strength contribution from the dislocation strengthening can be expressed as Equations (2) and (3) using a modified Taylor model [39–41]:

$$\sigma_D = \alpha GMb \sqrt{\rho} \tag{2}$$

$$\rho = \frac{3(1-f)}{bL}\bar{\theta}_{LABs} \tag{3}$$

where α is a constant taken as 0.24 [42], G is the shear modulus taken as 45.6 GPa [42], M is the Taylor factor taken as 3.06 [42], b is the Burger's vector taken as 0.256 nm [42], ρ is the dislocation density, f is the fraction of HABs, $\overline{\theta}_{LABs}$ is the mean misorientation angle of LABs and L is the mean boundary spacing. It should be noted that this work assumes all dislocations to be the LABs with the misorientation angles lower than 15° but larger than 2°. Thus, it neglects the LABs with misorientation angles less than 2° and the individual dislocations between LABs. There are two reasons. One is that the EBSD examination is related to the angular resolution besides the lateral and longitudinal resolutions. The angular resolution usually is about 1° when no particular care is imposed on the EBSD examination [43]. Thus, a safe and popular operation is to take 2° as the lower limit of the misorientation angle of LABs. The other is that the ECAP process is one of severe plastic deformation and the evolution of dislocation configures is faster than that of the conventional deformation method. For example, the mean misorientation angles of LABs in the central and bottom regions were more than 4° after only one ECAP pass. This indicates that a number of dislocations were stored in the distinguishable LABs whose misorientation angles were much more than 2°. In other words, the dislocations stored in the LABs with misorientation angles less than 2° and the individual dislocations between LABs were subordinate and negligible.

The strength contribution from the grain boundary strengthening can be expressed as Equation (4) using a modified Hall–Petch relationship [39–41]:

$$\sigma_{GB} = k\sqrt{f/d} \tag{4}$$

where k is a material parameter taken as 0.14 MPa·m$^{-1/2}$ [42], and d is the average grain size. It should be noted that there is an apparent difference between the modified and conventional Hall–Petch relationships, that is, the fraction of HABs (f) is introduced as the modified factor. It can be understood in this way. The ECAP process caused the well-developed dislocation structures where a number of LABs exist in the grains divided by HABs. Qualitatively, when dislocations attempt to slide in the grains and pile-up in the front of the grain boundaries, the LABs have impeded the motion of these dislocations. The strengthening mechanism in this case is different from that in the conventional case where the dislocations slide casually and only grain boundaries impede their motion. Thus, it needs a parameter to decrease the ability of the conventional grain boundary strengthening. Quantitatively, the grain radius (d/2) is inversely proportional to the grain boundary area per unit volume [40]. Thus, for the grain containing a number of LABs, it also needs to use the fraction of HABs to modify the grain boundary area per unit volume.

Figure 7 shows the calculated and experimental yield strengths in the central and bottom regions of the ECAPed Cu-Mg alloy, and Table 1 additionally shows their strength contributions from the various strengthening mechanisms. For the central region, the experimental yield strengths have been reported in the literature [15]. For the bottom region, it is impossible to perform a standard tensile testing due to the dimensional limit of the ECAPed billets, and thus only the calculated yield strengths are listed. Before the ECAP process, the Cu-Mg alloy shows a much lower experimental yield strength of only 124 MPa due to the dynamic recrystallized microstructure [15]. After one ECAP pass, the experimental yield strength increases significantly to 370 MPa. The calculated yield strength in the central region (344.7 MPa) shows that the improvement should be mainly attributed to the strength contribution from the dislocation strengthening (205.9 MPa) and the grain boundary strengthening (21.8 MPa). In addition, the calculated yield strength in the bottom region (291.4 MPa) is 53.3 MPa lower than that in the central region, showing a 16% difference where 46.6 MPa is from the dislocation strengthening and 6.7 MPa is from the grain boundary strengthening. Such a great strength difference is apparently resulted from the microstructure difference in the two regions. With the number of ECAP passes, the constantly accumulated dislocations are absorbed by the subgrain boundaries, leading to the increase in the misorientation angle and the formation of the HABs. Thus, the strength contributions from the dislocation strengthening and grain boundary strengthening enhance, resulting in continuous improvements in the yield strengths. Finally, after eight ECAP passes, the experimental yield strength achieves a high value of 548 MPa. The calculated yield strength in the central region (555.0 MPa) shows that the strength contributions from the dislocation strengthening (335.8 MPa) and grain boundary strengthening (102.2 MPa) are predominant, accounting for 78.9% in the whole. In addition, with the number of ECAP passes, the strength difference in the two regions gradually weakens and nearly disappears. For example, the difference decreases from 16% after one ECAP pass to 10% after two ECAP passes, and further decreases to 2–4% in the subsequent ECAP process.

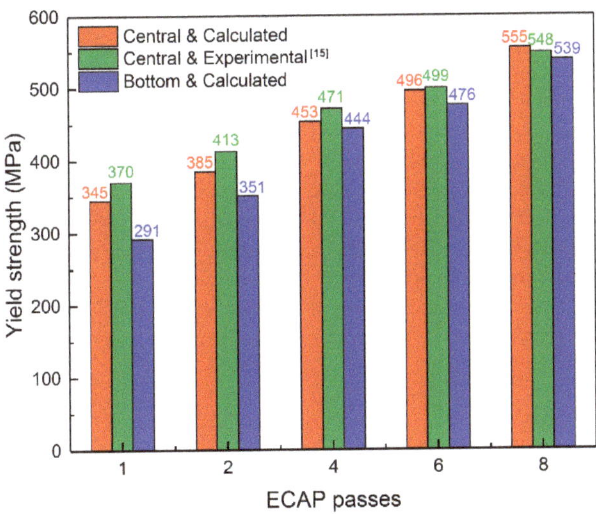

Figure 7. Calculated and experimental [15] yield strengths in the central and bottom regions of the ECAPed Cu-Mg alloy.

Table 1. The strength contributions from various strengthening mechanisms and the calculated and experimental yield strengths.

Region	ECAP Passes	σ_{LF} (MPa)	σ_{SS} (MPa)	σ_D (MPa)	σ_{GB} (MPa)	σ_y^{cal} (MPa)	σ_y^{exp} (MPa)
Central	1	68.3	48.7	205.9	21.8	344.7	370 [15]
	2	68.3	48.7	237.7	30.8	385.4	413 [15]
	4	68.3	48.7	280.8	55.5	453.3	471 [15]
	6	68.3	48.7	294.3	84.5	495.8	499 [15]
	8	68.3	48.7	335.8	102.2	555.0	548 [15]
Bottom	1	68.3	48.7	159.3	15.1	291.4	
	2	68.3	48.7	209.1	24.6	350.8	
	4	68.3	48.7	279.0	47.9	443.9	
	6	68.3	48.7	286.7	72.7	475.8	
	8	68.3	48.7	332.8	89.0	538.8	

An important point is that the effect of the dislocation strengthening is much more significant than the effect of the grain boundary strengthening according to the above-mentioned calculated result. This is different from the conventional view that the UFGed materials are mainly strengthened by the grain boundary strengthening. This question should depend on how to define the dislocation strengthening and the grain boundary strengthening. In this work, the grain boundaries are considered as HABs with misorientation angles larger than 15° and the dislocations are considered to be stored in LABs with the misorientation angles lower than 15° but larger than 2°. Some researchers hold the view that the boundaries with misorientation angles larger than 5° or 10° play a similar role in strengthening, just like the grain boundaries providing grain boundary strengthening. Thus, these boundaries are also generalized as grain boundaries. It results in the difference in considering the dislocation strengthening and the grain boundary strengthening for different research works. In addition, characterization methods also significantly affect how to distinguish the grain boundaries (or called HABs) and subgrain boundaries (or called LABs). For the EBSD examination, the misorientation angles of material elements are original data. How to define boundaries is dependent on researchers. For the TEM examination, however, the grain boundaries are mainly distinguished according to the contrast of the bright-field image. For example, Figures 4 and 5 in the reference [23] show some bright-field images of ECAPed pure Al and Al alloy, respectively. These so-called grains are adjacent to each other but show similar contrast. Apparently, the similar contrast means the similar orientations of these grains. Then, these grains actually are subgrains divided by LABs rather than grains divided by HABs. This also results in the overestimation of the grain boundary strengthening and the underestimation of the dislocation strengthening.

Figure 8 shows the relationship between the calculated yield strength and the measured hardness in the central and bottom regions of the ECAPed Cu-Mg alloy. However, the result is unclear. For the first six ECAP passes, the calculated yield strength and the measured hardness show a good linear relationship and high correlation coefficients more than 0.99. However, after eight ECAP passes, the correlation coefficients between the calculated yield strength and the measured hardness decrease to 0.96–0.97. From the variation trends of them, the increment of the calculated yield strength per the increment of the measured hardness increases sharply after eight ECAP passes. Thus, the linear relationship between the calculated yield strength and the measured hardness is predicted to be worse for more than eight ECAP passes.

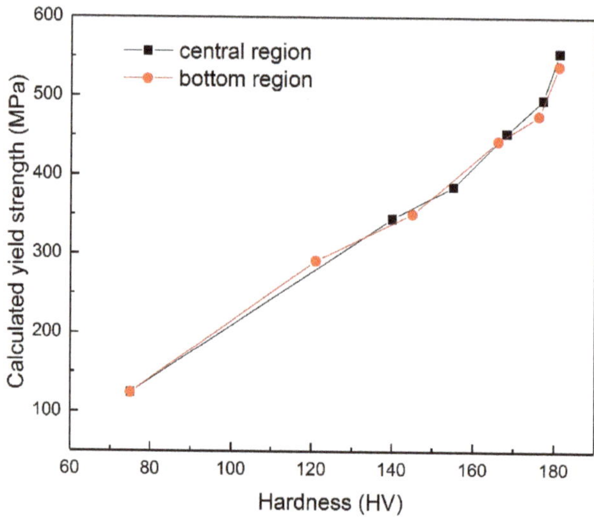

Figure 8. Relationship between the calculated yield strength and the measured hardness in the central and bottom regions of the ECAPed Cu-Mg alloy.

5. Conclusions

1. The hardness of a Cu-0.43Mg (wt.%) alloy was improved significantly at the initial stage of ECAP process, and the lower-hardness region appeared at the area nearby the bottom surface. With the number of ECAP passes, the hardness gently increased and finally became saturated. The inhomogeneity of the hardness distribution along the normal direction gradually weakened and finally disappeared;

2. The shear microstructure in the central region was different from that in the bottom region after one ECAP pass, and they became similar to each other after two ECAP passes, except the rotation angle of the elongated grains. With the further increasing ECAP passes, there was no obvious microstructure difference between the central and bottom regions;

3. The inhomogeneities of the hardness and the microstructure along the normal direction should be attributed to the non-zero outer arc of curvature of the ECAP die and the friction between the bottom surface of the billets and the ECAP die walls;

4. The strengthening mechanisms showed that the improvement of yield strengths should be mainly attributed to the strength contribution from the dislocation strengthening and the grain boundary strengthening. For example, after eight ECAP passes, they accounted for 78.9% of the yield strength of 555.0 MPa.

Author Contributions: Formal analysis, M.M. and X.Z.; Funding acquisition, Z.L., Z.X. (Zhu Xiao) and H.J.; Investigation, M.M., X.Z., H.H. and Z.X. (Zhu Xiao); Methodology, M.M.; Project administration, Z.L., Z.X. (Ziqi Xia) and H.J.; Resources, Z.L. and Z.X. (Zhu Xiao); Supervision, Z.L., Z.X. (Zhu Xiao) and H.J.; Validation, M.M. and X.Z.; Writing—original draft, M.M.; Writing—review and editing, Z.X. (Zhu Xiao). All authors have read and agreed to the published version of the manuscript.

Funding: This research was funded by the National Key Research and Development Program of China (Grant No. 2016YFB0301300), National Natural Science Foundation of China (Grant No. 51974375), Technology Research Program of Ningbo, China (Grant No. 2019B10088), and Project of State Key Laboratory of Powder Metallurgy, Central South University, Changsha, China.

Acknowledgments: Thanks are also given to Zhilei Zhao for his help.

Conflicts of Interest: The authors declare no conflict of interest.

References

1. Feng, J.; Liang, S.; Guo, X.; Zhang, Y.; Song, K. Electrical conductivity anisotropy of copper matrix composites reinforced with SiC whiskers. *Nanotechnol. Rev.* **2019**, *8*, 285–292. [CrossRef]

2. Zhao, Z.; Xiao, Z.; Li, Z.; Qiu, W.; Jiang, H.; Lei, Q.; Liu, Z.; Jiang, Y.; Zhang, S. Microstructure and properties of a Cu–Ni–Si–Co–Cr alloy with high strength and high conductivity. *Mater. Sci. Eng. A* **2019**, *759*, 396–403. [CrossRef]

3. Li, S.; Guo, X.; Zhang, S.; Feng, J.; Song, K.; Liang, S. Arc erosion behavior of TiB_2/Cu composites with single-scale and dual-scale TiB_2 particles. *Nanotechnol. Rev.* **2019**, *8*, 619–627. [CrossRef]

4. Feng, J.; Song, K.; Liang, S.; Guo, X.; Jiang, Y. Electrical wear of TiB_2 particle-reinforced Cu and Cu–Cr composites prepared by vacuum arc melting. *Vacuum* **2020**, *175*, 109295. [CrossRef]

5. Ji, G.; Li, Q.; Li, L. The kinetics of dynamic recrystallization of Cu–0.4Mg alloy. *Mater. Sci. Eng. A* **2013**, *586*, 197–203. [CrossRef]

6. Ji, G.; Yang, G.; Li, L.; Li, Q. Modeling Constitutive Relationship of Cu-0.4 Mg Alloy During Hot Deformation. *J. Mater. Eng. Perform.* **2014**, *23*, 1770–1779. [CrossRef]

7. Yuan, Y.; Dai, C.; Li, Z.; Yang, G.; Liu, Y.; Xiao, Z. Microstructure evolution of Cu–0.2Mg alloy during continuous extrusion process. *J. Mater. Res.* **2015**, *30*, 2783–2791. [CrossRef]

8. Yuan, Y.; Li, Z.; Xiao, Z.; Zhao, Z. Investigations on Voids Formation in Cu–Mg Alloy During Continuous Extrusion. *JOM* **2017**, *69*, 1696–1700. [CrossRef]

9. Zhen, G.; Kim, Y.; Haochuang, L.; Koo, J.-M.; Seok, C.-S.; Lee, K.; Kwon, S.-Y. Bending fatigue life evaluation of Cu-Mg alloy contact wire. *Int. J. Precis. Eng. Manuf.* **2014**, *15*, 1331–1335. [CrossRef]

10. Zhang, X.; Han, J.; Chen, L.; Zhou, B.; Xue, Y.; Jia, F. Effects of B and Y additions on the microstructure and properties of Cu–Mg–Te alloys. *J. Mater. Res.* **2013**, *28*, 2747–2752. [CrossRef]

11. Duan, Y.; Xu, G.; Tang, L.; Li, Z.; Yang, G. Microstructure and properties of the novel Cu–0.30Mg–0.05Ce alloy processed by equal channel angular pressing. *Mater. Sci. Eng. A* **2015**, *648*, 252–259. [CrossRef]

12. Ha, S.-H.; Yoon, Y.-O.; Lim, H.-K.; Kim, S.K. Effect of Heat Treatment on Microstructure and Hardness of Cu–Mg Alloy with a Trace of Ca. *J. Nanosci. Nanotechnol.* **2017**, *17*, 7820–7823. [CrossRef]

13. Li, Y.; Xiao, Z.; Li, Z.; Zhou, Z.; Yang, Z.; Lei, Q. Microstructure and properties of a novel Cu-Mg-Ca alloy with high strength and high electrical conductivity. *J. Alloy. Compd.* **2017**, *723*, 1162–1170. [CrossRef]

14. Freudenberger, J.; Kauffmann, A.; Klauß, H.; Marr, T.; Nenkov, K.; Sarma, V.; Schultz, L. Studies on recrystallization of single-phase copper alloys by resistance measurements. *Acta Mater.* **2010**, *58*, 2324–2329. [CrossRef]

15. Ma, M.; Li, Z.; Qiu, W.; Xiao, Z.; Zhao, Z.; Jiang, Y. Microstructure and properties of Cu–Mg-Ca alloy processed by equal channel angular pressing. *J. Alloy. Compd.* **2019**, *788*, 50–60. [CrossRef]

16. Valiev, R.Z.; Langdon, T.G. Principles of equal-channel angular pressing as a processing tool for grain refinement. *Prog. Mater. Sci.* **2006**, *51*, 881–981. [CrossRef]

17. Cobos, O.F.H.; Berríos-Ortiz, J.A.; Cabrera, J.M. Texture and fatigue behavior of ultrafine grained copper produced by ECAP. *Mater. Sci. Eng. A* **2014**, *609*, 273–282. [CrossRef]

18. Wei, W.; Wang, S.L.; Wei, K.X.; Alexandrov, I.V.; Du, Q.B.; Hu, J. Microstructure and tensile properties of Cu Al alloys processed by ECAP and rolling at cryogenic temperature. *J. Alloy. Compd.* **2016**, *678*, 506–510. [CrossRef]

19. Zaynullina, L.; Alexandrov, I.V.; Wei, W. Effect of the stacking fault energy on the mechanical properties of pure Cu and Cu-Al alloys subjected to severe plastic deformation. *MATEC Web Conf.* **2017**, *129*, 2032. [CrossRef]

20. Wongsa-Ngam, J.; Wen, H.; Langdon, T.G. Microstructural evolution in a Cu–Zr alloy processed by a combination of ECAP and HPT. *Mater. Sci. Eng. A* **2013**, *579*, 126–135. [CrossRef]

21. Mishnev, R.; Shakhova, I.; Belyakov, A.; Kaibyshev, R. Deformation microstructures, strengthening mechanisms, and electrical conductivity in a Cu–Cr–Zr alloy. *Mater. Sci. Eng. A* **2015**, *629*, 29–40. [CrossRef]

22. Iwahashi, Y.; Wang, J.; Horita, Z.; Nemoto, M.; Langdon, T.G. Principle of equal-channel angular pressing for the processing of ultra-fine grained materials. *Scr. Mater.* **1996**, *35*, 143–146. [CrossRef]

23. Xu, C.; Furukawa, M.; Horita, Z.; Langdon, T.G. The evolution of homogeneity and grain refinement during equal-channel angular pressing: A model for grain refinement in ECAP. *Mater. Sci. Eng. A* **2005**, *398*, 66–76. [CrossRef]

24. Xu, C.; Langdon, T.G. The development of hardness homogeneity in aluminum and an aluminum alloy processed by ECAP. *J. Mater. Sci.* **2007**, *42*, 1542–1550. [CrossRef]

25. Beyerlein, I.; Toth, L.S. Texture evolution in equal-channel angular extrusion. *Prog. Mater. Sci.* **2009**, *54*, 427–510. [CrossRef]

26. Prell, M.; Xu, C.; Langdon, T.G. The evolution of homogeneity on longitudinal sections during processing by ECAP. *Mater. Sci. Eng. A* **2008**, *480*, 449–455. [CrossRef]

27. Alhajeri, S.N.; Gao, N.; Langdon, T.G. Hardness homogeneity on longitudinal and transverse sections of an aluminum alloy processed by ECAP. *Mater. Sci. Eng. A* **2011**, *528*, 3833–3840. [CrossRef]

28. Reihanian, M.; Ebrahimi, R.; Moshksar, M.; Terada, D.; Tsuji, N. Microstructure quantification and correlation with flow stress of ultrafine grained commercially pure Al fabricated by equal channel angular pressing (ECAP). *Mater. Charact.* **2008**, *59*, 1312–1323. [CrossRef]

29. Ma, M.; Li, Z.; Qiu, W.; Xiao, Z.; Zhao, Z.; Jiang, Y.; Xia, Z.; Huang, H. Development of homogeneity in a Cu-Mg-Ca alloy processed by equal channel angular pressing. *J. Alloy. Compd.* **2020**, *820*, 153112. [CrossRef]

30. Kim, H.S. Evaluation of Strain Rate During Equal-channel Angular Pressing. *J. Mater. Res.* **2002**, *17*, 172–179. [CrossRef]

31. Kim, H.S. Finite element analysis of deformation behaviour of metals during equal channel multi-angular pressing. *Mater. Sci. Eng. A* **2002**, *328*, 317–323. [CrossRef]

32. Kim, H.S.; Seo, M.H.; Hong, S.I. Finite element analysis of equal channel angular pressing of strain rate sensitive metals. *J. Mater. Process. Technol.* **2002**, *130*, 497–503. [CrossRef]

33. Beyerlein, I.; Tomé, C. Analytical modeling of material flow in equal channel angular extrusion (ECAE). *Mater. Sci. Eng. A* **2004**, *380*, 171–190. [CrossRef]

34. Suh, J.-Y.; Kim, H.-S.; Park, J.-W.; Chang, J.-Y. Finite element analysis of material flow in equal channel angular pressing. *Scr. Mater.* **2001**, *44*, 677–681. [CrossRef]

35. Deng, G.; Lu, C.; Su, L.; Liu, X.; Tieu, A. Modeling texture evolution during ECAP of copper single crystal by crystal plasticity FEM. *Mater. Sci. Eng. A* **2012**, *534*, 68–74. [CrossRef]

36. Skrotzki, W.; Toth, L.S.; Klöden, B.; Brokmeier, H.-G.; Arruffat-Massion, R. Texture after ECAP of a cube-oriented Ni single crystal. *Acta Mater.* **2008**, *56*, 3439–3449. [CrossRef]

37. Yang, G.; Li, Z.; Yuan, Y.; Lei, Q. Microstructure, mechanical properties and electrical conductivity of Cu–0.3Mg–0.05Ce alloy processed by equal channel angular pressing and subsequent annealing. *J. Alloy. Compd.* **2015**, *640*, 347–354. [CrossRef]

38. Lei, Q.; Li, Z.; Gao, Y.; Peng, X.; Derby, B. Microstructure and mechanical properties of a high strength Cu-Ni-Si alloy treated by combined aging processes. *J. Alloy. Compd.* **2017**, *695*, 2413–2423. [CrossRef]

39. Luo, P.; McDonald, D.; Xu, W.; Palanisamy, S.; Dargusch, M.; Xia, K. A modified Hall–Petch relationship in ultrafine-grained titanium recycled from chips by equal channel angular pressing. *Scr. Mater.* **2012**, *66*, 785–788. [CrossRef]

40. Hansen, N. Hall–Petch relation and boundary strengthening. *Scr. Mater.* **2004**, *51*, 801–806. [CrossRef]

41. Hansen, N. Boundary strengthening in undeformed and deformed polycrystals. *Mater. Sci. Eng. A* **2005**, *409*, 39–45. [CrossRef]

42. Rodriguez-Calvillo, P.; Ferrer, N.; Cabrera, J.M. Analysis of microstructure and strengthening in CuMg alloys deformed by equal channel angular pressing. *J. Alloy. Compd.* **2015**, *626*, 340–348. [CrossRef]

43. Germain, L.; Kratsch, D.; Salib, M.; Gey, N. Identification of sub-grains and low angle boundaries beyond the angular resolution of EBSD maps. *Mater. Charact.* **2014**, *98*, 66–72. [CrossRef]

MDPI

St. Alban-Anlage 66

4052 Basel

Switzerland

Tel. +41 61 683 77 34

Fax +41 61 302 89 18

www.mdpi.com

Crystals Editorial Office

E-mail: crystals@mdpi.com

www.mdpi.com/journal/crystals